IMPROPER LIFE

CARY WOLFE, SERIES EDITOR

18 *Improper Life: Technology and Biopolitics from Heidegger to Agamben*
Timothy C. Campbell

17 *Surface Encounters: Thinking with Animals and Art*
Ron Broglio

16 *Against Ecological Sovereignty: Ethics, Biopolitics, and Saving the Natural World*
Mick Smith

(continued on page 191)

IMPROPER LIFE

Technology and Biopolitics from Heidegger to Agamben

TIMOTHY C. CAMPBELL

posthumanities 18

University of Minnesota Press
Minneapolis
London

Published by the University of Minnesota Press
111 Third Avenue South, Suite 290
Minneapolis, MN 55401-2520
http://www.upress.umn.edu

Library of Congress Cataloging-in-Publication Data

Campbell, Timothy C.
Improper life : technology and biopolitics from Heidegger to Agamben / Timothy C. Campbell.
p. cm. — (Posthumanities ; 18)
Includes bibliographical references (p.) and index.
ISBN 978-0-8166-7464-0 (hc : alk. paper) — ISBN 978-0-8166-7465-7 (pb : alk. paper)
1. Biopolitics. 2. Death—Political aspects. 3. Technology—Philosophy. I. Title.
JA80.C36 2011
320.01—dc23

2011017140

Printed in the United States of America on acid-free paper

The University of Minnesota is an equal-opportunity educator and employer.

18 17 16 15 14 13 12 11 10 9 8 7 6 5 4 3 2 1

CONTENTS

PREFACE

Bíos between Thanatos and Technē

AT A RECENT CONFERENCE on politics and life, one of the foremost American scholars of Foucault observed that the inflation surrounding the term *biopolitics* had reached truly pernicious proportions. There was something so totalizing about the term, he argued, something so unwieldy, that made it unfit as a paradigm for understanding the kinds of resistances required of subjects today. Given the current context of budget cuts and the elimination of humanities departments (fall 2010), for this theorist (and he wasn't alone in thinking so), biopolitics fails to offer enough ballast, or any ballast at all, in navigating the struggles that are upon us.

Such a judgment may appear inopportune as an opening for a book whose principal focus lies on the merits of biopolitical reflection, especially when the topic under consideration is the continuing intersection of technology in all its forms with life. Yet the questions raised about the value added of biopolitical reflection need to be taken seriously. Is there something about the nature of biopolitical thought today that makes it impossible to deploy affirmatively? Or more gravely, does biopolitical thought do the dirty, intellectual work of neoliberalism, offering little opposition to local threats, while focusing exclusively on matters of life and death at the level of species? If the answer is a murky no (or, for that matter, a murky yes), then our current understanding of biopolitics may, in fact, be too indebted to death—that we have less a biopolitics at our disposal than a thanatopolitics, one we employ at our own peril.

Doubting the potential resistance provided by biopolitical reflection offers, in fact, a starting point for the following study on the relation between *bíos* and technology, between *bíos* and *technē*.[1] One of the arguments I will be making throughout these pages is that the reason contemporary biopolitics devolves into thanatopolitics so seamlessly often has to do with an often unexplored relation between *technē* and thanatos that appears across the work of a number of the most important philosophers writing today in a biopolitical key. Indeed, my impression is that death gains the upper hand in biopolitical reflection precisely at the moment when questions of technology grow in importance, which raises a question: what is it about *technē* that calls forth thanatos in a context of life? Is there an aporia with regard to *technē* that repeatedly shifts discussions about biopolitics toward a horizon of death, regardless of the context? If so, then locating the aporia would be essential since it might suggest ways of recovering different perspectives on *technē*. Said differently, if one could find the point at which *bíos* begins its drift to thanatos, one could consider ways of reframing *technē* or, for that matter, *bíos* as well.

Here I have been helped enormously by the readings that a number of Italian theorists of biopolitics have made of Martin Heidegger's thought, especially Giorgio Agamben and Roberto Esposito. For Agamben and, more recently, Esposito, a crucial reframing occurs when both translate Heidegger's *eigentlich* and *uneigentlich* not in the language of (in)authenticity but in a different register of what is and is not considered (im)proper. Agamben did so early in *Stanzas,* when discussing the improper features of metaphor (though clearly the distinction between *proper* and *improper* also deeply informs many of his other works from the 1990s, especially *The Coming Community* and *Means without End*).[2] Esposito, too, draws on translations of *proper* and *improper* in his reading of the role of community in Heidegger's thought in *Communitas.*[3] For my part, I began to wonder if the proper–improper operator could also be extended to those pages I knew best from Heidegger, namely, *Parmenides,* as well as to those texts in which Heidegger puts forward his questions and answers about technology more generally as a way of unearthing a relation between life and *technē*.[4] The reader will find the results of that investigation in chapter 1, in which I read proper and improper writing in a number of seminal works from Heidegger to show how forms of writing quickly extend to or contaminate life. My conclusion is that here in the improper and proper distinction, a productive way can be found for locating the appearance of death in life, thanatos in *bíos,* thanatopolitics in biopolitics.

If, in chapter 1, I describe a biopolitical Heidegger avant la lettre, chapter 2 reads more as a chronicle of death foretold in contemporary Italian thought, especially in the more recent writings of Agamben and the recently translated works from Esposito. Without wanting to rehearse too much my argument there, the principal problem underpinning the chapter concerns the intensification of the thanatopolitical in Agamben's recent *The Kingdom and the Glory,* but equally in *The Sacrament of Language,* and even more in the pamphlet *What Is an Apparatus?*[5] Indeed, much of the chapter turns on the term *apparatus (dispositif)* and Agamben's willful conflation of that term with a Heideggerian notion of improper writing. In the second part of the chapter, I turn to Esposito's reading of the biopolitical in *Third Person* and his collection of essays, *Terms of the Political,* as well as an important text, "The Dispositif of the Person."[6] On my read, Esposito attempts to think through an impersonal possibility for life that would avoid the difficulties of following Heidegger's line of inquiry too closely (though, as I point out, other problems of a different nature ensue). The third chapter centers on the work of that other philosopher who, along with Agamben and Esposito, is today mining the veins of thanatos in life. The thought of Peter Sloterdijk forms the subject of the chapter, in particular, the recent translations of two texts.[7] Here, as was the case with Agamben and Esposito, the proper and improper produce a deeply thanatopolitical reading of everything from biotechnology to rage.

As important as these stops are in an itinerary of thanatopolitical thought today, the underlying objection of the conference scholar remains unanswered. If the biopolitical is riven by technology's inscription in thanatos, then how useful can it actually be for writing "a critical ontology of ourselves"?[8] In the final chapter, I turn to this question when imagining a practice of *bíos,* one that might be able to avoid *technē*'s seemingly inevitable enmeshment in death. Beginning with Foucault's *Security, Territory, Population,* and then turning to his later *Hermeneutics of the Subject,* I show how Foucault locates a possible genealogy of biopower in *bíos*'s capture by the self through the test.[9] Using that as a foundation, I sketch a practice of *bíos* through the categories of attention and play as a way of responding to Foucault's diagnosis of the self as deeply responsible for biopower today. Drawing on Sigmund Freud's writings on the drives as well as negativity, along with the writings of Maurice Merleau-Ponty, Gilles Deleuze and Félix Guattari, D. W. Winnicott, and Walter Benjamin, I argue that a practice of *bíos* in terms of attention and play would be better able to evade the problems of mastery that characterize so many of the accounts of *technē* under study here. I end on a hopeful note by imagining a possibility for a practice of play that, following Nietzsche's thinking of perspectivism, might have, as a consequence, a "planetary movement"—the flip side of a globalization thought only or primarily in thanatopolitical terms.

Many friends contributed to the writing of this book. The first chapter began as the basis for a weeklong series of seminars I gave at the Italian Institute for the Human Sciences in Naples in 2008. My thanks to the students there, whose comments and suggestions they will find on very nearly every page here. I tried out the second chapter as well as sections of the last chapter while I was a writer-in-residence at Birkbeck's Law School in summer 2010. My thanks to Peter Fitzpatrick and Patrick Hanafin for the kind invitation as well as to the cadre of students and faculty who joined me, in particular, Julia Chryssostalis, Nathan Moore, and José Bellido. They, too, will see their many suggestions generously spread across these pages. Closer to home, I presented parts of chapter 4 to the Society for the Humanities at Cornell University while a fellow there in 2009–10. My thanks to my fellow fellows as well as to the society's director, Timothy Murray, for their attentive readings. I also want to thank the following friends for their numerous kindnesses while I was writing: Roberto Esposito, Sergia Adamo, Ida Dominijanni, Cesare Casarino, Franco Berardi, Adam Sitze, Gregg Lambert, Kevin Attell, Adriana Cavarero, Rosi Braidotti, Catherine Malabou, Karen Pinkus, Mitchell Greenberg, Bruno Bosteels, Laurent Dubreuil, Simone Pinet, Kate Bloodgood, Ruth Mas, Marie-Claire Vallois, Richard Klein, David Ferris, Federico Fridman, and Lorenzo Fabbri. I also want to thank Cary Wolfe and Douglas Armato for what can only be described as their supreme "posthuman" patience while I was finishing the project.

Finally, this book is dedicated to Michela Baraldi, Alessandro Campbell, and Nicolas Campbell: every day, you teach me how to play.

1 DIVISIONS OF THE PROPER

Heidegger, Technology, and the Biopolitical

THAT THIS CHAPTER should open with the thought of Martin Heidegger in a context of the thanatopolitical is perhaps surprising. Yes, it's certainly true that Heidegger's thought continues to generate enormous attention—one need only consider the titles that appear every year dedicated to him[1]—but my impression is that few have attempted to set out the profound connections that join his thought to a larger drift of contemporary thought toward the thanatopolitical.[2] In the following pages, I want to sketch the path of that drift by examining two terms that appear across Heidegger's thought. The first is immediately recognizable, occupying, as it does, a central place in four texts—in the series of lectures Heidegger gave in 1942–43 that are collected in *Parmenides,* especially those sections dedicated to the (im)propriety of the hand: Heidegger's reading of "Homecoming/To Kindred Ones" from 1942, published in *Elucidations on Hölderlin's Poetry*; 1954's "The Question Concerning Technology"; and finally, his "Letter on Humanism" from 1947. The term in question is *technē* and its derivative technology.[3] The second term is nearly as familiar to contemporary readers, though it is rarely, if ever, named in connection with Heidegger's thought. I'm speaking of *biopolitics,* the seemingly never-ending inscription of biology in politics as well as the reverse: of politics read in a biological key.[4] As I consider the intersection of technology and biopolitics in Heidegger's later thought, one of my principal arguments will be precisely that to the degree we speak about biopolitics today, lurking beneath is a conception of technology deeply indebted to Heidegger's ontological elaboration of it. Moreover, this intersection provides the ground for the marked thanatopolitical inflection of biopolitics that characterizes so much of contemporary political philosophy. I will have much more to say about this in chapter 2, when I discuss the thought of Giorgio Agamben, especially his reflections on the function of *dispositif* in *oikonomia*. I also take it up in a somewhat different fashion in my reading of Roberto Esposito's understanding of *dispositif* and personhood, while in the third chapter, dedicated to Peter Sloterdijk, the figure (or, depending on one's point of view, the phantasm) of Heidegger dramatically reappears in Sloterdijk's immunological walls, the technology of the household, and most forcefully, the distinction between humanizing and bestializing media. As I argue there as well, how Heidegger takes up the question of technology, in the distinction between proper and improper writing, allows an implicit thanatopolitics to become available for contemporary political thought. In these four texts, Heidegger elaborates a distinction between proper and improper writing that has ontological effects such that a division in life is constructed between one *Art*, or species of man, associated with proper writing and another with improper writing.

One final observation on the theme of thanatopolitics, technology, and contemporary Italian thought. It's true that the object of thanatopolitical reflection in Agamben's work is chiefly Nazism. Indeed, Agamben refers in most cases to the presumed biological need (and practice) of making some live by killing others because the presence of these others can no longer be tolerated. This is how he will read modernity as populated entirely by *homo sacer*. Yet the ultimate premise for these readings can be found in the relation of technology to Being in Heidegger's thought, that is, in an ontological tear brought on by the distinction between proper and improper forms of writing. That Agamben deploys the state of exception as the mechanism by which biopolitics is always already a thanatopolitics doesn't alter, however, the fundamental authorization that Heidegger's thought provides Agamben because proper and improper writing frequently appear to be the basis for his distinction between forms of life. Agamben himself suggests just such a reading in his "Notes on Politics" from *Means without End*, in which he extends *proper* and *improper* into a global critique of industrial democracy.[5] In Agamben's positing of a relation between technology and sacrifice that is embodied in the figure of the *homo sacer*, he implicitly gestures to a Heideggerian ontology that would have technology determining what is proper or not proper to man. Similarly, Roberto Esposito has made Heidegger's thinking of technology the object of a reflection on the origin of politics (and the impolitical) both in *Communitas: The Origin and Destiny of Community* and *L'origine della politica: Simone Weil o Hannah*

Arendt?, both of which might appear to the reader to be on the way to the thanatopolitical. Admittedly, this is much less the case with *Bíos* and *Third Person,* in which his more recent considerations on Nietzsche are intended to provide a counterweight to the thanatopolitical valence that the encounter with Heidegger seems inevitably to produce.[6] My own reflections here are intended more as symptomology of the tragic possibilities technology provides the thanatopolitical.[7] In other words, I don't want to circumscribe thanatopolitics simply to Nazism but rather to see it as implicit whenever Being, language, or life is divided against itself into proper or improper.

Technology and the Propriety of the Hand

In no text more so than *Parmenides,* Heidegger's lecture from the winter semester of 1942–43, does one find the parameters of life and technology so intimately bound together. Of particular importance are those sections dedicated to the typewriter, in which Heidegger posits a fundamental ontological distinction between a proper writing, a *Festschrift* or "handwriting," and another thought against (and through) the example of the typewriter. These pages are well known in the field of media studies and informed much of late 1990s work originating in Germany that attempted to develop a systems approach to media.[8] Clearly a closer look at this text is necessary if we are to get at the lurking category of the thanatopolitical.

Heidegger begins by asking what distinguishes the hand that writes from the hand that types. His answer is that man acts "for the hand is, together with the word, the essential distinction of man."[9] Indeed, the hand, for Heidegger, provides the essential difference between man and animal: "No animal has a hand, and a hand never originates from a paw or a claw or talon."[10] This biotechnical distinction between hand and paw cannot be thought apart from the word: the hand is coterminus with the word as well because "the hand sprang forth only out of the word and together with the word," as he will go on to say.[11] This simultaneous move of precedence and superimposition is typical of Heidegger's thought, especially later in "The Question Concerning Technology," in which he posits the origin of politics in a form of technology that precedes and is the cause of the political.[12] In *Parmenides,* too, the question of technology is joined to the hand that writes—handwriting—in another related figure that cannot be thought apart from it: inscription. It is inscription that will form such an important part of the discourse of authenticity that characterizes Heidegger's thought (as well as Adorno's well-known critique of it).[13]

Yet a contrasting figure quickly emerges in Heidegger's defense of inscription. Heidegger, writing by hand, calls it *dictation*:

> It is not accidental that modern man writes "with" the typewriter and "dictates" (the same word as "poetize" [*Dichten*]) "into" a machine. This "history" of the kinds of writing is one of the main reasons for the increasing destruction of the word. The latter no longer comes and goes by means of the writing hand, the properly acting hand, but by means of the mechanical forces it releases. The typewriter tears writing from the essential realm of the hand, i.e. the realm of the word. The word itself turns into something "typed." Where typewriting, on the contrary, is only a transcription and serves to preserve the writing, or turns into print something already written, there it has a proper, though limited significance. . . . Mechanical writing deprives the hand of its rank in the realm of the written word and degrades the word to a means of communication. In addition, mechanical writing provides "this advantage," that it conceals the handwriting and thereby the character. The typewriter makes everyone look the same.[14]

The passage is dense in its medialogical implications as well as in what it portends for the relation of thanatopolitics and technology. I want to consider, in particular, three questions Heidegger raises.

First, note that Heidegger distinguishes between the hand that writes and the hand that types, and then sees the latter as having emerged out of the hand through its mechanical imprinting. His objective, as he will go on to argue, is to disclose a former, more originating relation of Being to handwriting: "In handwriting the relation of Being [*des Seins*] to man, namely the word, is inscribed [*eingezeichnet*] in beings themselves."[15] Consider how Heidegger lines up the word with man—man doesn't simply embody the word but *is* the word to the degree he writes with his hand. This assumption allows Heidegger to argue that the hand modern man has is no longer "the properly acting hand" when the hand types and does not write. The effect of this transformation is violent: proper writing is torn away from the hand by the very same forces released in fact by the hand. This suggests that Heidegger wants to mark a proper action for the hand and an improper one associated with the typewriter. The proper one is precisely the one that inscribes Being in man, whereas the improper relation between the hand and the typewriter concerns "something typed," something "that is only a transcription and serves to preserve writing." This distinction between proper and improper writing will soon be extended from handwriting–typewriting to the practice of hermeneutics itself. It's with this in mind that Heidegger's final directive must be understood, namely, that his collected works not appear in a critical edition but as writings *aus letzter Hand,* which is to say, "the volumes in the series come 'straight from his hand' and contain a minimum of scholarly apparatus."[16]

Next, consider that the context for Heidegger's entire discussion of proper and improper writing is anchored to the appearance of typewriting as a form of technology and a mode of dictation in *modernity.* Indeed, that the ultimate context of Heidegger's reflections concerns modernity is never in doubt. He refers to modern man as he who writes "with" the typewriter, and he will say soon after that the object of his analysis is "modern man." If we consider other forms of technology contemporary with the typewriter, the implicit connections with the improper writing enacted by the typewriter will be confirmed as well. The gramophone, as Friedrich Kittler observes (but also the telephone and, as I have argued elsewhere, wireless telegraphy, with its headsets, writing hands, and signed communiqués), also makes dramatic use of a writing liberated from the ontological norms of handwriting.[17] In other words, it is in the modern period that improper writing becomes the norm, with all the attendant normalizing consequences on Being.

These normalizing effects are decisive for elaborating a biopolitical perspective on Heidegger's thought. The profound affiliation he sees between modern forms of technology and their capacity to occlude handwriting, and with it, the character of the individual who writes by hand, alters the relation of being to Being. This becomes clear in the closing line of the preceding passage, when Heidegger inscribes proper and improper writing within the political horizon in which all men are made the same. Or better, the condition for communication in the modern period will be precisely this move from proper to improper, where proper connotes not simply a hand that writes but what belongs to man as properly his own.[18]

Implicit, therefore, in the distinction Heidegger draws in *Parmenides* between proper and improper forms of writing is a relation of man to his writing mechanisms (and "not really machines," as Heidegger notes in the strict sense of machine technology). He is arguing that with these changes, political consequences naturally follow in that all men are made the same when the typewriter dominates Being. It is this tension between a proper relation with Being that man enjoys when he writes by hand and another, degraded form of improper writing that moves Heidegger's discourse toward another question, toward what destabilizes that which belongs most properly to man, namely, his relation to Being as expressed in an action that is his own (the writing hand). In other words, Heidegger places the identity of the properly acting hand in opposition to another form that not only puts at risk identity as such but, more dramatically, endangers a proper relation to Being.

The third question follows quickly on these and concerns the name we will want to give this other form of improper writing. For Heidegger, the term will be *communication.* Here, though, *communication* must be read in conjunction with its etymological roots in *community,* the *co-munus* or shared *munus,* a form of gift giving that, as Roberto Esposito demonstrates, cannot be thought apart from the demands the *co-munus* makes on individual identity itself.[19] This is not the place to map all the deep connections running between Heidegger's thinking of community and technology; rather, I want to observe that in the stark difference Heidegger posits between a writing that preserves a relation of Being to man and another that puts it at risk, we find ourselves witness to one of the most important political and idolatrous figures of Heidegger's thought of technology. In this difference between writing in the modern period, we find an implicit alliance between improper writing born of technology and a political form in which the identity of the one who writes is put at risk. The idolatrous nature of improper writing is that it awards a power to the collective capable of persuading men and women that they more properly belong to a collective.[20]

We can see this alliance more clearly in the discussion Heidegger offers soon after his reading of the typewriter when he turns to Leninism and its metaphysics:

> The bourgeois world has not seen and in part still does not want to see today that in "Leninism," as Stalin calls this metaphysics, a metaphysical projection has been performed, on the basis of which in a certain way the metaphysical passion of today's Russians for technology first becomes intelligible, and out of which the technical world is brought into power.[21]

What goes unexplained in Heidegger's account, however, is precisely what accounts for this complete technical organization of the world in Bolshevism. Lurking beneath Heidegger's analysis of this technical world is an unspoken connection between Leninist metaphysics and the improper act of writing. And although he doesn't explain why Bolshevism more than other forms of metaphysics experiences technology so "unconditionally and radically," we may assume that the answer will be found in the greater com-munal pressure that is exerted on individual identity in Leninism—that the degradation of the relation of Being to man is greater where all are made the same.[22] Heidegger's anticommunism thus goes hand in hand with his wide-ranging critique of improper writing, both founded on a perceived anxiety related to threats to man's proper relation to Being.

It's here that a tear in Being emerges, forcing Heidegger's ontology to drift toward the tragic and thanatopolitical. We see it if we focus on the nature of the subject of technology, which is to say, on what kind of human being is it who masters technology and, relatedly, what price he pays for such mastery. In answer to the first, Heidegger writes,

> Perhaps the much discussed question of whether technology makes man its slave or whether man will be able to be the master of technology is already a superficial question, because no one remembers to ask what kind of man [*welche Art Mensche*] is alone capable of carrying out the "mastery" of technology. The "philosophies" of technology pretend as if "technology" and "man" were two "masses" [*Grössen*] and things simply on hand.[23]

What kind of man masters technology?[24] The change in the species of man that attempts to extend his domination over technology—and we note that Heidegger rarely, if ever, employs the word *use (Gebrauch)* in this context—is in fact what is most dangerous about technology. A dramatic change in the nature of humanity itself arises from this encounter of man and technology, one that Donna Haraway, in a heroic moment of postmodernism, will call the cyborg, or what Hal Foster, glossing Freud, will call "prosthetic gods."[25] Less present in these conscious postmodern appropriations of Heidegger, though, is Heidegger's full-blown anxiety about the change in species that man undergoes in the attempt to dominate technology. Furthermore, for Haraway and Foster, clearly there is no ideological animus toward communism as there is in Heidegger, where, in a context in which technology threatens to change humanity, it is in Bolshevik Russia that a radical world of technology appears to have emerged almost fully formed. The threat of technology, for Heidegger, apparently cannot be thought apart from its affiliation with attacks on what distinguishes one man from another, which is to say that the ultimate aim of Heidegger's reading is to defend a certain *Art* of man from the encounter with technology. This is the orientation that we ought to give his bringing down to size the "masses" of mankind as literally a thing that is hermetically sealed from the outside, that merely acts on (improperly) objects with the aid of technology.

In the distinction between mankind as a mass and mankind as species, it is technology that creates a tear in Being. Or differently, in man's attempt to master technology, it becomes possible to see on what basis a certain view of mankind as a bounded and protected entity depends on denying any sort of movement between the one who masters technology and the masses bounded on all sides from the untoward effects of technology. It becomes possible to distinguish between kinds of men and women, depending on the relation they enjoy with technology. On one side are those who continue to maintain a proper relation to Being, that is, to their own proper action when writing, and on the other, those others who, in mastering technology, have been altered such that they become a "kind of man" [*welche Art Mensche*].

Furthermore, in this distinction, an implicit value is already given to one kind of man over another because of the context in which Heidegger elaborates it: on one side are those who are all made equal, and on the other is the individual who, by writing properly, enjoys a relation to Being that the former do not. Much remains to be said about this caesura that Heidegger posits. It is a moment that also figures prominently in so many contemporary discussions of thanatopolitics, especially of the kind that Agamben offers in *The Open* as well as his more recent *What Is an Apparatus?* Yet it is also important to observe how Heidegger identifies the nature of the man who still may enjoy a relation to Being through a proper writing. Attempting to determine the features of technology, he writes, "Insight into the 'metaphysical' essence of technology is for us historically necessary if the essence of Western historical man is to be saved [*gerettet bleiben soll*]."[26] A number of points need to be made straightaway. First, for Heidegger, the other kind of man, one who exists in opposition to the standardized subject and object of technology, is Western, historical man. The descriptor *Western* here takes on more weight when we note that its clear contrary in the "Third Directive" of *Parmenides* is the technologized Leninist metaphysician; again

Heidegger's anticommunism bubbles along the surface of his reflections. Second, consider that Western man's relation to technology is one related to Western man's essence. It recalls, of course, Heidegger's later "Letter on Humanism," in which he asks, "But in what does the humanity of man consist?," which precedes a brief description of Marx's views on man's humanity as "recognized and acknowledged in society."[27] Here, though, the essence in question is one limited to Western man.

Finally, Heidegger endangers the essence of this man who is in need of being saved. Technology not only operates by creating a tear in Being and so in positing kinds of men, one who attempts to master and is made the same as everyone else and another (which is precisely why the typewriter works so elegantly as a stand-in for all forms of improper writing). In so doing, Heidegger places Western man, that is, the one who continues to write properly, in danger. The danger to Western man is intensified, as well, by the other descriptor here used as a marker, namely, *historical.* Indeed, in the same paragraph, Heidegger connotes history in metaphysical terms as presenting a danger: "He who has ears to hear, i.e. to grasp the metaphysical foundations and abysses of history and to take them seriously *as* metaphysical, could already hear two decades ago the word of Lenin: Bolshevism is Soviet power + electrification."[28] Heidegger's Western man constantly teeters over the abysses of history, given that "technology is entrenched in our history [*Die Technik ist in unserer Geschichte*]."[29] Implied is the threat of falling out of Being into a merely "technical world" of the Leninist sort. It is only with the elaboration of improper writing mechanisms that we recognize the monumental stakes involved in obtaining insight into "the metaphysical essence of technology."

Where do we see the drift toward thanatopolitics in Heidegger? We see it in his positing of Western man as requiring saving from the pernicious effects of technology on man's proper relation to Being. Said somewhat differently, Heidegger has essentially created a new grouping where before only mankind or humanity existed, configuring two forms of life where before there was only one who acted (and wrote) properly; the first is given over entirely to the social, technical world of metaphysical passion, whereas the other, Western man, is he who must be saved. This will be a crucial reflection for contemporary philosophy since Heidegger's distinction between proper and improper already puts Western man at risk. To use technology means already to be dominated by it in such a way that one loses what is most proper to mankind, namely, a relation to Being. The sense here is that when one is made the subject of improper writing, then a life is created by division whose value is lessened given that it is made the equal of everyone else. A divide comes to separate Western historical man, who sits astride the precipices of history—and therefore who must be saved, given the imminent danger—and those others who, in attempting to master technology, become its subject. Heidegger will use the term *modern* to name these subjects of technology [*der moderne Mensch*] who can still be saved as well as those completely mastered by technology. It is important to observe, however, that depending on their relation to technology, the points where these figures meet can in fact be moved, for instance, by political parties. Just such a possibility is implicit in Heidegger's reading of Bolshevism as "the 'organic,' i.e., organized, calculating (and as +) conclusion of the unconditional power of the party along with complete technicization."[30] This suggests that technology isn't simply immanent to itself but rather that institutions, a political party, or even a state can promote it.

It is this moment of configuring Western man as an essence that requires saving that will inform Giorgio Agamben's own reading of *homo sacer* and technology and about which I have much to say in the following chapter. At this point, I simply want to observe how Agamben's own reading of technology radicalizes the consequences of the moment when "Being [*das Sein*] has withdrawn itself from man and modern man has been plunged into an eminent oblivion of Being [*eine ausgezeichnete Seinsvergessenheit*]"; Agamben actualizes oblivion of Being in his reading as the creation of a life that can be taken with impunity.[31] Again, Heidegger authorizes this kind of reading to the degree that he sees the question of technology as one involving Western man's very essence. So, too, does Agamben's dehistoricization of *homo sacer* as emerging out of an infinite state of exception parallel Heidegger's view that "technology understood as modern, i.e. as the technology of power machines, is itself already a *consequence* and not the *foundation* of a transformation of the relation of Being to man [*des Bezugs des Seins zum Menschen*]."[32] I'll leave these other resemblances between Heidegger and Agamben for the next chapter, but I do want to suggest that one of the most important tributaries of contemporary political philosophy, namely, the eruption of life within the political, may be traced to these pages written in 1942–43. It is the eruption of death in saving life that comes to mark a form of life that has lost or has forgotten how a relation to Being through proper writing can be acted (on). Thanatos and the willful sacrifice of those who no longer enjoy a proper relation to the hand, who indeed no longer have a hand but instead merely manipulate (and in turn are manipulated)—are gestured to as well as authorized by Heidegger's discussion of the typewriter in *Parmenides*.

Note, as well, something else that Heidegger doesn't state explicitly, namely, that the greater or lesser use of technology can in fact be used by states or political parties to create a situation in which it becomes easier to take the lives of those from whom Being has withdrawn. The question to be posed at this juncture is precisely what forms of technology more successfully put in play the withdrawal of Being, or in the words of Friedrich Kittler, what forms of technology help to short-circuit the defenses of the human being from agreeing to be killed.[33] This impolitical question is implicit again in Heidegger's reading of Bolshevism, in which electrification and "Bolshevik power" account both for the completely technicized world and, with it, the sheer numbers of those who can be killed.[34] Ultimately, Heidegger's reading of proper and improper writing provides us with a paradigm with which to understand how improper writing deanchors Being, creating a form of life enthralled to technology to such a degree that it loses its "individual" features. These become clearer in those important pages dedicated to revelation and writing in "The Question Concerning Technology." It is to that text that I now want to turn.

Life and the Protocols of Writing

In "The Question Concerning Technology," Heidegger returns to the metaphysics of technology when he attempts to think *technē* again through a relation to proper and improper writing. Here, too, technology is profoundly inflected toward the thanatopolitical, especially toward another life inscribed in the register of those from whom Being has withdrawn. Yet, rather than choosing the typewriter as an example of impropriety, Heidegger instead moves toward a broader examination of technology, the general features of which have grown more intense in the succeeding years. The reasons we know well enough: the destruction of World War II as well as the birth of atomic terror. The most salient part of Heidegger's analysis concerns those sections in which he links the notion of revelation to that of *Stellen* and, with it, "a challenging-forth."

What results is a superimposition of improper writing with a revealing that moves through a series of challengings [*Herausfordern*]. Heidegger writes,

> The revealing that rules throughout modern technology has the character of a setting-upon [*Stellen*], in the sense of a challenging-forth. That challenging happens in that the energy concealed in nature is unlocked, what is unlocked is transformed, what is transformed is stored up, what is stored up is, in turn, distributed, and what is distributed is switched about ever anew. Unlocking, transforming, storing, distributing, and switching about are ways of revealing. But the revealing never simply comes to an end. Neither does it run off into the indeterminate. The revealing reveals to itself its own manifoldly interlocking paths, through regulating their course.[35]

Heidegger makes homologous the improper nature of technology with that of revelation, choosing to focus on the series of steps that technology employs for energy to be "challenged" into appearing. In a sense, Heidegger moves away from the question of the hand in *Parmenides.* He is still concerned with impropriety, but in "The Question Concerning Technology," it is improper writing that is shown to involve no simple mode of disclosing but, in fact, a number of moments that appear to be successive but are not. These steps—unlocking, transforming, storing, distributing, and switching—Heidegger explicitly calls ways of revealing. Thus he alters the framework of his earlier "meditation on unconcealedness" in the modern period to a more impolitical reflection on the means by which revelation and modern technology work together. On the one side, in *Parmenides,* we find a recapitulation of how modern technology may be thought through its relation to the unconcealedness of Being. In "The Question Concerning Technology," a modern technological regime emerges that is so powerful that no one-dimensional notion of revelation can be localized; rather, we find modes of revelation that build on earlier moments of disclosure.

Much has been made, of course, of these various moments of disclosing. Indeed, an entire field of media studies based around the notion of media as ecology is deeply indebted to the implicit reading of media as homologous to modes of revealing.[36] But I want to put forward another, perhaps complementary reading of revealing in Heidegger that shifts the register of these reflections toward the thanatopolitical. In this perspective, the unconcealment of technology, which enjoys its own particular form of revealing, never fully arrives, which is to say that implicit in the difference between the unconcealedness of writing that is proper to man and revelation that doesn't properly belong to him, we find the idea that the modes by which revelation are enacted never come to an end. In "The Question Concerning Technology," so-called modern man is made over into the subject of these modes of revealing, which is to say that modern man becomes the subject of a infinite loop of disclosing. He becomes so when placed in the position of a technological object; the telephone operator, the wireless operator, the secretary who takes dictation, the manager who dictates: each of them is a subject of modern technology that puts them in the position of those who need to be saved. Given that the human component is deeply involved at every level of revelation in modern technology, the subject of never-ending revelation continually appears.

In the preceding passage, Heidegger emphasizes modern technology rather than simply modern man, as he does in *Parmenides*. The change in focus does not mean, however, that modern man has simply been replaced by the tremendous power of modern technology. Modern man will be found if we look closely at those sections in which he introduces another term to mark the modern mode of revealing, namely, *Bestand* or "standing reserve":

> What kind of unconcealment is it, then, that is peculiar to that which comes to stand forth through this setting-upon that challenges? Everywhere, everything is ordered to stand by, to be immediately at hand, indeed to stand there just so that it may be on call for further ordering [*Bestellung*]. Whatever is ordered about in this way has its own standing [*Stand*]. We call it standing-reserve [*Bestand*]. The word expresses here something more, and something more essential, than mere "stock." The name "standing-reserve" assumes the rank of an inclusive rubric. It designates nothing less than the way in which everything presences that is wrought upon by the challenging revealing. Whatever stands by in the sense of standing-reserve no longer stands over against us as object.[37]

Heidegger doesn't simply limit the question of technology to objects but rather suggests that the question of technology is one intimately linked to man. Furthermore, the possibility that everything everywhere is ordered to stand by also includes modern man. Western man is placed in the position of those who require saving, given the improper relation of the hand to transcription. The process by which man is threatened, or better, the process whereby Being is no longer inscribed, takes place through an unconcealment, a notion of revelation that repeats itself endlessly, not just in the act of writing down but also in what improper writing allows us to see for the first time: writing broken into five protocols—unlocking, transforming, storing, distributing, and switching—that remain unrevealed in a proper writing of inscription.

The question of mankind, and of the kind of life he leads and lives, is posed directly in this improper and impolitical act of writing that encompasses Being. The question is one of being at hand—a hand that waits in the ready for a future ordering and a future revelation. Furthermore, it is this being at hand that characterizes the metaphysical underpinning of technology. Of course, Heidegger doesn't simply leave it there; instead, he turns again to the relation he had already drawn between modern man and technology in *Parmenides*. Standing reserve [*Bestand*] will now name the mode by which everything is brought into relation with revelation, and there can be no doubt that the reference also encompasses man himself. In different words, Heidegger's critique of technology in "The Question of Technology" doubles back to his earlier critique of Western man, now shorn of the descriptor *Western*, to offer a representation of the subject of technology as homologous to a subject of revelation. To be subject of this disclosing means to remain in place, to be at hand, in a "challenging claim which gathers man thither to order the self-revealing as standing reserve: '*Ge-stell*' [Enframing]," as Heidegger will go on to say.[38] The crucial point arrives soon after, when Heidegger once again pushes up against another abyss—not of history but destining and the danger it represents for man.

Note, too, how in the passage, man is placed in a certain position vis-à-vis technology by technology itself. The effects are deleterious. Two possibilities for revelation are allowed. The first is characterized by an "ordering," which Heidegger previously had associated with "standing reserve" [*Bestand*]. The second possibility involves a more originary relation of revelation to man's own proper essence. Between these two possibilities, Heidegger anticipates a danger for man as "he comes to the very brink of a precipitous fall."[39] In this condition, man "is endangered from out of destining. The destining of revealing is as such, in every one of its modes, and therefore, necessarily, *danger*."[40] The problem concerns rightly the nature of what is disclosed by technology as not concerning man at all, not even as object, but rather—and this is the crucial point for a genealogy of thanatopolitics and technology in contemporary philosophy—as standing reserve. If that is the case, then "man in the midst of objectlessness is nothing but the orderer of the standing-reserve."[41]

We can express this somewhat differently. Technology exposes us to the danger that we are no longer the subject of our own proper unconcealing (that we are a mere object or slave of technology, hence the "objectlessness" of the position). Rather man is destined to wait as standing reserve, as one "ordered" by technology. This Heidegger confirms immediately after: "But Enframing does not simply endanger man in his relationship to himself and to everything that is. As a destining, it banishes man into that kind of revealing which is an ordering."[42] This "ordering" places man in the position of being at hand, in a position vis-à-vis technology, in which what is most proper to him, namely, "to enter into a more original revealing":[43]

> The essence of technology lies in Enframing. Its holding sway belongs within destining. Since destining at any given time starts man on a way of revealing, man, thus under way, is continually approaching the brink of the possibility of pursuing and pushing forward nothing but what is revealed in ordering, and of deriving all his standards on this basis. Through this the other possibility is blocked, that man might be admitted more and sooner and ever more primally to the essence of that which is unconcealed and to its unconcealment, in order that he might experience as his essence his needed [*gebrauchte*] belonging to revealing. Enframing belongs within the destining of revealing.[44]

Much can be said about this moment when man is exposed to the danger of revealing induced by technology. Consider, first, the importance of the distinction between proper and improper, which, though less manifest than in *Parmenides,* becomes the axis around which the danger of technology is to be understood in "The Question Concerning Technology." The revealing associated with technology is improper to the degree that it puts man at risk in his essence. In *Parmenides,* the essence in question belonged to Western, historical man, whereas here it is man in general, reflecting a dramatic extension of modern technology's capacity to hijack Being.[45] Technology imposes itself on man but also positions him in relation to it so as to transform him into standing reserve, into an object of revelation. Put in this position, man is enframed *(gestellt)* but also endangered, for the same reasons that an improper writing endangered him in *Parmenides*: what is proper, namely, the relation to a unconcealing of truth, has now been made more distant. Distancing marks man as the subject of improper revelation and is therefore what endangers him.

What does this danger consist of? Heidegger never articulates the threat outside the terms *essence, ordering,* and fostering the *saving power,* the latter of which, of course, evokes the earlier figure of Western, historical man.[46] There the threat was posed by the typewriter, which threatened mankind with the loss of identity. Here the threat is more abstract, filtered through the essential relation to a more originary relation to revelation. Yet the key terms in both instances are saving as well as being destined for (*Ge-schick* and *schicksal*). This is where so much of contemporary Continental philosophy will be found, which is to say that technology places mankind in a position in which it is destined to be "ordered" for the future needs of technology. What some will do, Agamben principally, is to find in this moment of being a ceaseless command to be on call for technology: life itself has somehow been diminished such that it becomes easier to take the lives of those who have been "ordered" by technology. The text presents a number of opportunities for such a reading: from Heidegger's repeated references to "human willing" when speaking of freedom and, in conjunction with that, the suggested inhumanity of a technological standing reserve [*Bestand*].[47] We find as well the sense of man as somehow banished from the realm of proper writing and revelation into another in phrases such as "as a destining, it banishes man into that kind of revealing which is an ordering."[48] This suggests, in turn, that those who look like everyone else from the point of view of standing reserve and improper revelation are the ones most endangered by technological ordering. Agamben's great insight among many—and these others I'll turn to in the following chapter—is to have seen in this endangering not simply the need, should we choose to call it that, for saving those who are ordered but, more profoundly, the possibility for taking their lives on a scale that few had previously grasped. In Heidegger's breakup of revelation into "unlocking, transforming, storing, distributing, and switching," the power of technology's "ordering" has been expanded to such a degree that man himself can be killed more easily. It is both a small and monumental step from there to Agamben's reading, in which the object of technology's "ordering" can be enrolled among *homines sacri.*

At this juncture, Heidegger introduces another possibility near the end of "The Question Concerning Technology" through the well-known quote of Hölderlin's: "*Wo aber Gefahr ist, wächst / Das Rettende auch*" ("But where danger is, grows / The saving power also"). The possibility of a saving power that grows with danger is one that Agamben will read impolitically as proof of the power of thanatopolitics in the contemporary period. If we want, it is the Heidegger of *Parmenides* who, in a certain sense, comes to dominate the Heidegger of "The Question Concerning Technology." The saving power that Heidegger posits in the latter becomes, in Agamben's notion of the remnant, an enormous multiplier of the danger that modern technology represents for mankind given the reciprocal inscription of being saved and being marked for death thanks to the operation of the state of exception. I'll have much more to say in the following chapter, but in the meantime, let me suggest as a way of pivoting to another of Heidegger's texts that figures so prominently in contemporary theorizations of the thanatopolitical that this saving power is one that also informs Roberto Esposito's own reflections on the impersonal; indeed, when Esposito speaks of reversing a Nazi thanatopolitics into a biopolitics of life through a philosophy of the impersonal in *Bíos* and *Third Person,* he, too, works out of a Heideggerian conception of the thanatopolitical. Yet Esposito, for his part, does not radicalize this saving power as negative; instead, he will note the idolatrous critique of technology that is implicit throughout "The Question Concerning Technology." In short, Esposito will read the saving power not as one *over* life (as Agamben does) but as one *of* life that encompasses all forms of life.

Mystery, Technology, Proximity

Over the previous pages, I've surveyed some of the most important moments of drift in Heidegger's thought toward the thanatopolitical, instances that emerge out of Heidegger's deep association of improper writing in *Parmenides* and a technologically rendered revelation that places mankind at risk. In the following section, I want to turn to two other texts that together provide more detail on the thanatopolitical effects of technology while confirming the divisions the technology creates between those who enjoy a proper and improper relation to revelation and writing. The first comes from Heidegger's "Homecoming," an essay from 1941–42, collected in *Elucidations of Hölderlin's Poetry.* Here Heidegger writes at length both of the notion of homeland, a theme to which he returns in "Letter on Humanism," and the nature of poetic speech. What interests me is how Heidegger thinks the poet's vocation to "homecoming" as a form of technology, not as an act of writing but rather through the compression of distance, or better, as the oscillation between distance and "nearness" that comes to characterize those who have a homeland. Heidegger lays out the details by noting how "nearness" to the homeland described in Hölderlin's poem "Heimkunft" operates through the maternal voice:

> Suevien dwells near the origin. This nearness is mentioned twice. The homeland itself dwells near. It is the place of nearness to the hearth and to the origin. Suevien, the mother's voice, points toward the essence of the fatherland. It is in this nearness to the origin that the neighborhood to the most joyful is grounded. What is most characteristic of the homeland, what is best in it, consists solely in its being this nearness to the origin—and nothing else besides it.[49]

We note again how the hand appears as the mode by which the maternal voice indicates what is most essential and innermost in the homeland. In pointing to this more essential fatherland, the maternal voice suspends man in "nearness" to the homeland. The poetic word, availing itself of the maternal voice, becomes the means by which "a nearness which still holds something back in reserve" is constructed.[50] This notion of "nearness," which has informed so much of media theory over the last twenty years, gestures both to what brings the near nearer and at the same time what makes it distant, since what is sought after cannot be near in Heidegger's judgment.[51] The result is an understanding of distance as both compression and extension:

> The nearness that now prevails lets what is near be near, and yet at the same time lets it remain what is sought, and thus not near. We usually understand nearness as the smallest possible measurement of the distance between two places. Now, on the contrary, the essence of nearness appears to be that it brings that which is near, yet keeping it at a distance.[52]

In other words, Heidegger broadens (or circumscribes, depending on one's point of view) the definition of *nearness* to the degree it is seen not simply as distance or nearness but as the continual movement between them. The result? "This nearness to the origin is a mystery."[53]

The introduction of mystery as an effect of the movement between distance and nearness allows us to inscribe Heidegger's reading of the poetic word in a larger context of technology and thanatopolitics.[54] The poet, by appropriating the maternal voice, inhabits what Heidegger calls "a nearness which still holds something back in reserve."[55] So doing, the poet draws listeners toward the homeland. Quoting Hölderlin's "Heimkunft"—"Cares like these, whether he likes them or not, a singer / Must bear in his soul, and often, but the others not"—Heidegger distinguishes negatively between those who hear and those who do not: "the others."[56] These others are not yet enrolled as belonging to the homeland, though given the nature of the mystery of the nearness, "which still holds something back in reserve," the possibility exists that these others, too, may be made to draw closer.

The seemingly different terminology that Heidegger adapts here doesn't prevent us from seeing the overlap between a kind of word that creates "nearness" and another term that he will later use to describe the workings of technology, namely, *unconcealedness*. The word will appear in the closing paragraphs of "The Question Concerning Technology," in which Heidegger again employs mystery in naming another oscillation that structures so much of his thinking on technology. There he writes,

> The irresistibility of ordering and the restraint of the saving power draw past each other like the paths of two stars in the course of the heavens. But precisely this, their passing by, is the hidden side [*das Verborgene*] of their nearness. When we look into the ambiguous essence of technology, we behold the constellation, the stellar course of the mystery.[57]

When read together with the earlier passage, in which Heidegger thinks of nearness as a mystery, we can see how Heidegger superimposes technology and the poetic word over nearness; whether the operator of that nearness is the poetic word or technology as its improper form, they both traffic in mystery. In other words, the mystery of the poetic word in its "nearness which still holds something back in reserve" is superimposed over the mystery of technology to the degree that in the nearness between "ordering" and "the restraint of the saving power," there, too, its mystery is charted.

In chapter 3, I argue that this superimposition of technology and the mystery of the poetic word is the not-so-hidden center of Peter Sloterdijk's essay "Rules for the Human Zoo: A Response to the *Letter on Humanism,*" for there the poetic word becomes the source of "proper" or humanizing nearness with technology as its spectral and improper other.[58] But as long as we inscribe the poetic word and technology in the horizon of media, as most media theoreticians and Sloterdijk do, we will fail to appreciate the originality of much of recent Italian thought, which poses the question of life through a reading of "The Question Concerning Technology" and indirectly in *Elucidations of Hölderlin's Poetry.* By invoking the movement between "the restraint of the saving power" and "ordering," we see that life is once again at stake. Indeed, in "The Question Concerning Technology" primarily, it is the life of the one "ordered" that is present in the "revealing that in the technological age rather conceals than shows itself," which is to say, she who is ordered to wait for an ultimate disclosing that never fully arrives.[59] The same holds true for *Elucidations of Hölderlin's Poetry,* except that here Heidegger lodges the saving power, never directly named as such, within the heart of a proper "nearness which still holds something in reserve." This saving power is never distant from the homeland for Heidegger, and yet the position of the hearer vis-à-vis the source still entails a mystery. The only difference would appear to be that this listener who draws nearer to homeland is not governed by the saving power of technology.

The consequences of such a reading are clear. The potential thanatopolitical drift of how Heidegger considers technology cannot be thought outside a mystery that is seen as more proper or, if we want, as one that doesn't call forth this saving power. There is no moment in *Elucidations of Hölderlin's Poetry* in which the poetic word orders life in ways that make it necessary for a saving power to function thanatopolitically. The poetic word, by drawing the listener closer to the homeland, isn't seen as endangering her (and hence requiring safekeeping); rather, the implicit suggestion is that those who hear and draw near thanks to the poetic word are in a different relation to the truth. With that said, a subterranean biopolitical moment also lurks precisely in the difference between those who draw near and those who cannot. Such a moment fundamentally concerns the thought of the community in Heidegger. Esposito's exposition of this moment merits our attention:

> The community is therefore recognized according to its originary essence, its originary "having been." This is the terrible syllogism that captures Heidegger from within his own discourse . . . that transforms the "in-common" of everyone into *a* particular community intent on conquering a proper future through a rediscovery of the purest origin. This—and nothing else—was Heidegger's Nazism: the attempt to turn himself directly towards the proper, to separate it from the improper, to make it speak in the affirmative, originary voice.[60]

What Esposito sees as operating fundamentally in the thought of Heidegger is precisely the distinction between the in-common of everyone and the in-common of a particular community, which then is manifested across other areas of Heidegger's thought, most decisively in the German community as opposed to the one that includes all of mankind, or for that matter, Western historical man as opposed to another nonhistorical, "Eastern" man. In other words, the biopolitics of Heidegger's thinking doesn't reside merely in the arising of a saving power in conjunction with technology's ordering but rather will be found earlier, in a distinction between one's own and what is not, whether it be a proper community or not, or pace Agamben's later, thanatopolitical reading of People as a "whole and integral body politic" and people as a "fragmentary multiplicity of needy and excluded bodies."[61]

Another consequence emerges from this genealogy of technology and biopolitics, one that concerns more generally the superimposition between the poetic word and technology. Where the poetic word offered the possibility of transforming "the others" into proper listeners or readers (and so enrolling them as those who "know" the essence of the homeland), so, too, does technology draw them closer to the mystery. This capacity to draw near entails a biopolitical component. My own work in the history of wireless telegraphy and, later, voice transmission confirms as much. Indeed, one of the principal effects of radio will be to dramatically increase the numbers of listeners who are drawn closer to the "mystery" of the homeland. The point, of course, of this biopolitical practice was to increase biopower by turning directly to what was considered to belong properly to the nation or state, namely, listeners. Thus the Nazis literally addressed what they saw as the proper Being of Germany to promote biopolitical effects. Radio not only intensified biopower to the degree that the numbers of listeners grew but also helped bring about a technologically inflected people, a people of listeners seduced by the mystery of "nearness which still holds something back in reserve." Given this initiation into wireless hearing, a more intense biopolitical entity was formed, a technologized *Volk* who could be addressed as one in an instant—a more strident body politic to the degree that this body politic was brought ever closer to the source of the mystery. Obviously, the effect of this drawing closer, this impolitical knowledge of the homeland, is to make their sacrifice all the easier.[62]

Note that not only the Nazis and Italian fascists put this impolitical knowledge, with its breaking apart, in the modern era of the protocols of proper writing and their emerging biopolitical effects into practice. Liberal democracies, who were intent on stopping them and protecting themselves from aggression, also did. In the case of radio transmissions, for instance, the classic image is, of course, Hitler speaking in front of a cluster of microphones and then a cut to a large number of people listening raptly to his words, only to break into thunderous applause soon after. It is also true, however, that Franklin Roosevelt had his fireside chats, and then, of course, the radio offered Winston Churchill a medium by which those who listened to his words could be transformed into a form of life that could be more easily sacrificed. Churchill's repeated references, for instance, in his speech of June 4, 1940, to "island" and "empire" signaled fundamentally to his listeners what was properly British.[63] In this case, too, a technologically inflected people was constructed in a hurry to meet the emergency at hand. In each case, however, the biopolitical effects of technology are felt decisively in short-circuiting the proper defenses of the individual vis-à-vis the larger body politic. Subjects are created who are willing to die, in the first instance, to protect the German *ghènos*, and in the second, to defend against aggression by deploying one of the most powerful modes for increasing the biopower at the "liberal" state's disposal.

With that said, in no way do I want to level the political or, for that matter, ethical differences between the Nazi use of the radio and British or American use. With the Nazis (though it is equally true for Mussolini), whenever the radio is spoken of, one finds obvious Nietzschean overtones—what Ernst Nolte in a not-so-different context refers to as a "metaphysics of the glorification of life," which is then translated into a metaphysics of the technological voice or, better, a glorification of a certain form of life that is valued "spiritually" to the degree that it speaks over the radio.[64] In the case of the Nazis, only those forms of life that speak and move accordingly are spiritualized; anyone else is banished into nonexistence: it is a small step from the spiritualization of those who speak because they move to the spiritualization of race in Nazi biopolitics. The role that radio played in furthering the spiritualization of these forms of life ready to be sacrificed is one that deserves much more attention than has been given to it up to now. In any case, a corollary spiritualization of the British or American listener cannot be said to have taken place.

More work is needed on the thanatopolitical features of the radiophonic transmission, just as one would want to distinguish between these forms of heightened biopower during the war and those associated with liberal democracies after the war. Here the thanatopolitical effects will be less marked not only because the war is over (and another, the cold war, had just begun, which would require only intermittent biopolitical use of technology, given that the ultimate thanatopolitical moment resides in nuclear war) but also because technology will now be co-opted for a (bio)politics of the individual, so as merely to meet, or so it would seem, his material needs. What doesn't change, however, is that in both instances, power is deployed on bodies in what we will want to describe, along with Foucault, as a power over life, or what Heidegger might call the acceleration of the two stars passing each other in the heavens: "ordering" and "that which saves."

"Letter on Humanism": Biopolitics

The biopolitics of the individual in liberal democracies: is it inevitably thanatopolitical? That is one of the principal questions I want to ask in the pages that remain by looking at one of Heidegger's most important statements on the subject: "Letter on Humanism." The text is a familiar one, and so there's no need to linger over the specific details that gave rise to it; rather, I want to focus on three moments—let's call them *threats*—that appear in the letter.[65] The first occurs after Heidegger has unambiguously inscribed his earlier discussions of technology within a larger horizon of philosophy such that philosophy emerges as a slipping out of thinking into technology. The result is that "philosophy becomes a technique for explaining from highest causes."[66] This happens for the same reason that proper writing breaks up into different modes of writing and revealing: where Being has withdrawn from man in *Parmenides,* here, too, Being withdraws, or to paraphrase Heidegger, comes less to preside over thinking. It is this withdrawal of Being never directly recounted as such in the letter that provides the ground for Heidegger's relentless critique of humanism. He writes,

> In competition with one another, such occupations publicly offer themselves as "isms" and try to offer more than the others. The dominance of such terms is not accidental. It rests above all in the modern age upon the peculiar dictatorship of the public realm. However, so-called "private existence" is not really essential, that is to say free, human being. It simply insists on negating the public realm. It remains an offshoot that depends upon the public and nourishes itself by mere withdrawal from it. Hence it testifies, against its own will, to its subservience to the public realm. But because it stems from the dominance of subjectivity the public realm itself is the metaphysically conditioned establishment and authorization of the openness of individual beings in their unconditional objectification.[67]

Distinguishing between private "existence" and "the peculiar dictatorship of the public realm," Heidegger superimposes earlier historical Western man over "private existence" and "the technical organization of the world" of Bolshevism over "dictatorship of the public realm," which will include both the Soviet Union and Western liberal democracies (authorized by the grouping of all *-isms* under the banner of philosophy and its "occupations," where the echo with the former "ordering" can still be heard). Whereas before, Western historical man could still be saved from the onslaught of another dictatorship of the public realm (Leninism), here no distinction is made between Leninism and, let's call it for the time being, liberalism; instead, from Heidegger's impolitical perspective, both *-isms* are mere "occupations" that cover over a larger orientation "of the modern age," in which individual beings without distinction are forced to open to their "objectification." The threat that put Western, historical man in the position of requiring saving has apparently reached such an extreme that with the withdrawal of Being, a new being emerges, whose "openness" is no longer conditioned by the respective *-ism* in question.

Obviously, the withdrawal of private existence and its homonym *individual being* from the dictatorship of the public realm deserves scrutiny. For instance, the distinction Heidegger draws between the "free, human being" and the "individual being" gives pause because it would appear to create two "beings," one a spectral version of the other, given that mere "private existence" has meaning only thanks to its withdrawal from the public realm (whereas a "free, human being" would not in fact withdraw because its relation to Being has in no way been transformed). It is, of course, a classic moment of the exclusionary inclusion or the including exclusion that we have come to know so well from Agamben. It also accounts for one of the central understandings of the impolitical as well as immunity, as put forward by Esposito, in which a certain form of life attempts to immunize itself from a communal dominance of subjectivity.[68] Yet what stands out most is the reference to openness, which the reciprocal inscription of individual being in public dictatorship brings about.

What does *openness* refer to then? Heidegger doesn't answer directly, but perhaps the question is badly put. Openness for Heidegger isn't followed by an objective phrase, as in "openness to," but rather functions genitively: "in their unconditional objectification." To be open is to be unconditionally objectified. Why we discover in the following sentences: "Language thereby falls into the service of expediting communication along routes where objectification—the uniform accessibility of everything to everyone—branches out and disregards all limits. In this way language comes under the dictatorship of the public realm which decides in advance what is intelligible and what must be rejected as unintelligible."[69] The point of being open is to create conditions for accessibility so that communication may be expedited. Once again, one will hear echoes between "uniform accessibility" and the foregrounded anxiety that Heidegger felt in *Parmenides* toward the uniformity that the typewriter instituted, which is as it should be given the similar contexts between communication here and improper writing there. The threat is expressed similarly, as well, on the next page: "The widely and rapidly spreading devastation of language not only undermines aesthetic and moral responsibility in every use of language; it arises from a threat to the essence of humanity."[70]

The nature of the threat, not surprisingly, is both biological and political, which is to say that those individual beings who are open to communication become both the subjects and objects of a power over life at the moment when "language surrenders itself to our mere willing and trafficking as an instrument of domination over beings. Beings themselves appear as actualities in the interaction of cause and effect."[71] Unlike in previous texts, the threat to humanity arises at the same moment when beings can be dominated thanks to a transformation in language and its relation to Being. The specific nature of the threat in *Parmenides* and "The Question Concerning Technology" wasn't framed, though, in terms of language but instead in terms of writing and revelation. In "Letter on Humanism," Heidegger folds those concerns into a more global question of language to come to terms with the monumental theme of humanism. In other words, communication goes hand in hand with the greater possibility of domination over beings.

The second moment follows directly from these reflections and concerns the division that will mark the remainder of Heidegger's discussion. Turning to the context of care for man, Heidegger sees a gap between man and humanity: "Where else does 'care' tend but in the direction of bringing man back to his essence? What else does that in turn betoken but that man *(homo)* becomes human *(humanus)*? Thus *humanitas* really does remain the concern of such thinking. For this is humanism: meditating and caring, that man be human and not inhumane, 'inhuman,' that is outside his essence."[72] Note the division between man and human here. By distinguishing between them, Heidegger essentially has made the human, man's proper form, while man himself becomes essentially an improper form—improper in the sense that the man who remains merely man remains distant from Being and from the litany of associations that Being has for Heidegger, truth, in particular.[73] Here in all its glory, then, is the thanatopolitical division between *homo* and *humanus*. Why thanatopolitical? Because Heidegger has created a fissure between those who are properly human and those who are not. Once the tear has been made complete, man cannot be thought except within and outside his own missing or supplemental humanity.

Clearly, for Heidegger, the reason for such a division is to be able to institute the notion of care as a possible response to the "nearness which still holds something back in reserve" that he spoke of in *Elucidations of Hölderlin's Poetry*.[74] This "nearness which still holds something back in reserve," as we recall, left man enthralled to an improper revelation that never created increasing proximity to Being. Care, instead, will mark the path that the properly human will set out for the merely man. Much of the remainder of the "Letter on Humanism" turns precisely around the figure of this in-human, or, perhaps better, un-human figure and how he may draw near Being and so become *humanus*. But the question arises, if this figure isn't *humanus*, then how are we to go about describing him? For Heidegger, this not yet fully human figure can only be thought through the problematic metaphysical category of *animalitas*:

> We can proceed in that way; we can in such fashion locate man within Being as one being among others. We will thereby always be able to state something correct about man. But we must be clear on this point, that when we do this we abandon man to the essential realm of animalitas even if we do not equate him with beasts but attribute a specific difference to him. In principle we are still thinking of homo animalis—even when anima [soul] is posited as animus sive mens [spirit or mind], and this in turn is later posited as subject, person, or spirit [Geist]. Such positing is the manner of metaphysics. But then the essence of man is too little heeded and not thought in its origin, the essential provenance that is always the essential future for historical mankind. Metaphysics thinks of man on the basis of animalitas and does not think in the direction of humanitas.[75]

Heidegger's argument doubles back to an earlier moment when he posited a difference between a free, human being and an individual being, made so by a devastation of language. Here, though, Heidegger doesn't deny that man as an individual being is made to appear next to other beings without distinction. Such a consideration does have the advantage of being correct, but in saying so, an abyss opens up underneath the as yet not fully human figure. Heidegger appears to be arguing that in no way can one award *animalitas* to man without discounting any future difference that one might want to find that will distinguish man from a beast. Once an animal, always an animal, despite any later designations that grow out of soul that would include person or subject. We note that Heidegger, in observing this difference (that makes no difference), does not disown his earlier reflections on man and humanity. What he has done instead is to provide further ammunition in his salvo against a *humanitas* that continues to be inscribed in a metaphysical horizon of *animalitas*.

The thanatopolitical emerges from the folds of these reflections when Heidegger does not reject the basic distinction between *animalitas* and *humanitas*. It is clear why he does not because his own discourse is intent on working out from the inside of metaphysics to a firmer foundation for deconstructing *humanitas*. But the point is that by agreeing that man can continue to be distinguished from *humanus*, Heidegger still assumes a metaphysical distinction as the basis for moving toward an essential *humanitas*.[76] Despite these gyrations between animal and human, there remains a stark division between the human and another, potentially human, if provided with sufficient care. Inscribing "historical mankind" in some "essential future" does not in any way save mankind from the consequences of marking some or most as animals and others as more properly human.

The essential distinction between *animalitas* and *humanitas* adds an important element to our reflections. Implicitly raised is the possibility that technology contributes to this underlying and essential distinction between man as *animalitas* and man in some future as *humanitas*. To the degree that technology functions to bring about "a nearness which still holds something back in reserve" and not an essential nearness to Being, it holds man in place vis-à-vis an essential future that never arrives. So, too, do Heidegger's reflections on the improper writing of the typewriter as creating a uniformity allow us to see more deeply what it is that characterizes the *animalitas* of "Letter on Humanism." For Heidegger, a communication that makes everyone the same is to be thought now through the optic of *animalitas*, which in turn suggests not only a movement toward the essentially human but equally a movement toward an essential *animalitas*. Indeed "The Question Concerning Technology" as well as *Parmenides* may be read exactly in this way, as reflections on how *humanus* degrades into an improper form of the human thanks to technology. In "Letter on Humanism," Heidegger will then attribute to this other form of man the label of *animalitas,* despite the vacillation that characterizes so much of the middle section of the letter. When taken together, the previous texts under examination provide us with a fuller picture of how technology might be linked to man's dehumanization, while in "Letter on Humanism," Heidegger shows how this dehumanization is to be thought through the category of *animalitas.* And here it bears repeating that for Heidegger, the full title of person or subject in no way signals any true move in the direction of a future *humanus.*

For Heidegger, man is endangered if he remains on either side of the divide between *humanitas* and *animalitas.* It is true that Heidegger never directly describes the nature of the danger—in these texts, it is always spoken of in terms of essence, or the withdrawal of Being, or the oblivion of concealing. Nor does Heidegger explicitly pronounce that the inhuman man may be killed with impunity, nor, of course, that the properly human is premised on the death of these animal men, as Nazism assumes. Yet we can say that the Heideggerian ontology of Being presupposes the lesser form of the human, in a division that today, thanks to Agamben, we refer to as *zoē* and *bíos.* If today the thanatopolitical seems to dominate contemporary perspectives on biopolitics as well as our understanding of neoliberalism and globalization, it is because of this deep ambiguity concerning man and technology and the dehumanizing effects the latter has for man (whether we locate it, as Agamben does, in some transhistorical past or, as in Esposito's case, as emerging with the dawn of modern immunization in Hobbes).

The Nearness of Thanatos: Improper Writing Today

By way of conclusion, let me set out what I see, then, as some of the most important features of the drift toward the thanatopolitical in Heidegger's thought. The most important concern the effects of technology on the possibility of mankind achieving some sort of essential relation with Being, which then leads to a fundamental distinction between those, on one side, who are mere subjects of communication; those who later will be enrolled among the ranks of an *animalitas*; and others who, thanks to a proper writing, are seen as free, individual human beings, capable of "care," and never as mere subjects or persons. What results is an implicit discounting the former as less than human.

Equally, the distinction Heidegger draws on more than one occasion between individual being and a free, human being becomes a paradigm for a number of philosophers writing in a biopolitical key today. Agamben will often substitute singularity for the human; Sloterdijk will argue that power has shifted inextricably to individual as opposed to communal forms of life. For Sloterdijk, in fact, the care of which Heidegger speaks in "Letter on Humanism" has given way to nothing less than securing the individual through collective and noncommunal entities. This securing is directly at odds with the kind of securing that Heidegger speaks of in *Parmenides,* one that occurs thanks to the hand: "Where the essential is secured in an essential way, we therefore say it is 'in good hands,' even if handles and manipulations are not exactly necessary."[77] It also accounts for the wide-ranging critique Sloterdijk will launch against "insurance" in *Sphären* and elsewhere. Then we find Esposito, who attempts to meet the challenges of Heidegger's thought, which is to say, to address the implicit thanatopolitics of the individual being subjected to improper writing and revelation, by proposing a possible alternative through the impersonal. All, as I will have occasion to show, if not directly working out of Heidegger's own categories, do assume a sort of thanatopolitical tonality there.

I would add a final point as a way of bridging to the following chapter, and that concerns another distinction which I perhaps haven't foregrounded enough here, one that concerns the proper and improper distance that results from the two forms of writing and revelation that Heidegger speaks of in *Parmenides,* but especially in *Elucidations on Hölderlin's Poetry.* There we remember that Heidegger makes a distinction between nearness to Being and another, the technologically inflected "nearness which still holds something back in reserve." In contemporary readings of biopolitics, that "nearness which still holds something back in reserve" will be declined frequently as the principal definition of communication, that is, as enacting both a profound linguistic alienation of man and, perhaps even more important, as installing via technology a separation among men and women in which proper care cannot emerge. In communication, as in Enframing, "man stands so decisively in attendance on the challenging-forth of Enframing that he does not apprehend Enframing as a claim, that he fails to see himself as the one spoken to."[78] What emerges, therefore, is the decisive role technology plays in creating a scenario of danger and threat through separation—a tragic reading of technology providing the material necessary for contemporary perspectives on thanatopolitics to take form.

And yet this separation of man from himself equally entails a separation of man from man, depending on the order of observation adopted. What appears to be man's separation from himself when shifted to the perspective of his relation to others becomes "one final delusion: It seems as though man everywhere and always encounters only himself. . . . *In truth, however, precisely nowhere does man today any longer encounter himself, i.e. his essence.*"[79] In other words, thanks to technology, man remains separated from the possibility of meeting himself in others (or another) because man cannot make out any figure other than himself (which is precisely how technology works impolitically to block an opening to man himself through others). This separation of man is enacted through separation with others, a technological proximity that installs separation with man's own proper essence thought through the difficulty of man distinguishing himself from others. Separation makes domination over Being and beings possible, when the proximity that separates is associated with language.

This final reading of the notion of separation in Heidegger's thought that makes domination over Being possible leads us to the notion of *dispositif* in Giorgio Agamben's and Roberto Esposito's thought. What this chapter's itinerary through Heidegger's critique of technology demonstrates, however, is how the extension of improper writing tools makes mankind resemble others (and the other). In the move toward equality enacted through separation (and technology as primarily a mode for separating and not a practice), care for oneself and others is made difficult such that mankind is endangered. Heidegger's thanatopolitics will be found here, in the knowledge that where technology is augmented, (human) beings can be dominated. The conclusion can only be that separation is deeply connected to domination. Yet this raises almost more questions than it answers. How, for instance, can care be strengthened in a milieu of greater separation brought on by the expansion of improper writing machines? What accounts for the greater thanatopolitical effects of contemporary technology in Agamben's reading? Some of the answers will be found in the superimposition of Heidegger's improper writing machines over Foucault's notion of *dispositif.* It's to that deeply thanatopolitical move that I now turn.

2 THE *DISPOSITIFS* OF THANATOPOLITICS

Improper Writing and Life

IN THE PRECEDING CHAPTER, I took up the question of the relation between thanatopolitics and technology in the thought of Martin Heidegger by focusing on the distinction between proper and improper writing. In this chapter, I want to turn to two of the most important Italian philosophers writing today in an ostensibly thanatopolitical key: Giorgio Agamben and Roberto Esposito. To say that Agamben's thought is deeply indebted to Heidegger is, of course, to state nothing new. From the 1977 *Stanzas: Word and Phantasm in Western Culture* to his most recent *The Kingdom and the Glory* as well as *The Signature of All Things: On Method* and *The Sacrament of Language: Archaeology of the Oath*, Agamben's thinking of everything from infancy to the notion of paradigm grows out of a profound knowledge and abiding synthesis of Heidegger's work.[1] This is made explicit in a text like *The Open* even (or especially when) Agamben reads against Heidegger in those pages dedicated to the animal, the human, and boredom.[2] The following discussion, however, is not motivated by a possible anxiety of influence of Heidegger operating in Agamben's work because such a reading would limit the purview of Agamben's thought and its importance today for political reflection, turning our discussion toward the intertextual references with Heidegger that dot almost all the texts under examination here. As intriguing as that project might be, it would leave Agamben's thought inscribed solely within Heidegger's ontology.[3] Roberto Esposito, for his part, avoids many of the pitfalls of incorporating Heidegger's critique of technology into his own reading of the immunization paradigm, though that, too, comes with a price to be paid in resignation to what Heidegger calls "everydayness."

Therefore two questions will guide this chapter: first, what are the principal means by which Agamben and Esposito, when thinking the relation between biopolitics and technology, radicalize and extend the implicit Heideggerian division between proper and improper writing so as to configure their own vision of the thanatopolitical; and second, how is the notion of *dispositif* (apparatus), which plays such a prominent role in some of Agamben's more recent texts, to be thought in relation to this implicit thanatopolitics that emerges out of his appropriation and extension of Heidegger?[4] My impression is that the form of life that can be taken with impunity—*zoē* or *homo sacer* in Agamben's terminology—cannot be thought apart from a concern with technology, which is to say, cannot help but be inscribed in that horizon of improper writing and revelation that goes under the name of Enframing *(Gestellen).*

The Lexicon of Life

Any discussion of Agamben and Heidegger surely must begin with the opening distinction that Agamben makes between *bíos* and *zoē* in *Means without End*. Ours will be no different:

> The Greeks had no single term to express what we mean by the word "life." They used two terms that, although traceable to a common etymological root, are semantically and morphologically distinct: *zoē*, which expressed the simple fact of living common to all living (animals, men, or gods) and *bíos*, which indicated the form or way of life proper to an individual or group. In modern languages this opposition has gradually disappeared from the lexicon . . . one term only—the opacity of which increases in proportion to the sacralization of its referent—designates that naked presupposed element that it is always possible to isolate in each of the numerous forms of life.[5]

In this oft-quoted incipit, we find the basis for Agamben's thanatopolitical reading of technology. It will be premised, like so much of Agamben's understanding of contemporary politics, on a fissure in life itself, one that is coterminus with the birth of metaphysics. On one side resides *bíos*, which Agamben reads as a form of living that is immanent to itself, on the other side, the well-known—perhaps too well known, given the inflation of late—form of life known as *zoē*. The descriptor proper is one that we ought to focus on, not only because it is one of the principal ways

Heidegger thinks technology in *Parmenides* but also because it spells out the difference between a proper form of revelation associated with Being and another associated with technology. In the preceding passage, what qualifies for Agamben as a living properly belonging to a group is what distinguishes *bíos* from *zoē*. Such a reading is confirmed in a later chapter, "Notes on Politics." There Agamben writes,

> Praxis and political reflection are operating today exclusively within the dialectic of proper and improper—a dialectic in which either the improper extends its own rule everywhere, thanks to an unrestrainable will to falsification and consumption (as it happens in industrialized democracies), or the proper demands the exclusion of any impropriety (as it happens in integralist and totalitarian states).[6]

The passage seems straightforward enough. In addition to inscribing *bíos* and *zoē* in the broader horizon of proper and improper, Agamben indicates by *zoē* the improper form of living that characterizes "industrialized democracies," whereas *bíos* is situated in those totalitarian or fundamentalist states whose task it is to exclude the improper from the mode of living that is proper to the State.[7] Agamben's use of *totalitarian* here is clearly intended to bring together Nazi Germany and the Soviet Union in a way that recalls not so much Hannah Arendt but more those polemical pages of Foucault's "*Society Must Be Defended,*" in which Foucault argues for a biopolitics that would encompass both Socialism and Nazism through the category of racism as a form of thanatopolitics.[8] The least we can say about this passage from 1992 is that Agamben offers what will be a fundamental insight into contemporary political philosophy, not simply by offering a Foucauldian angle on how politics captures life but rather by discussing how the improper extends itself over proper political life in industrial democracies, while the proper conducts a campaign against the improper in totalitarianism. The subject of the improper as well as the proper for Agamben, however, remains life divided against itself.

I've glossed this passage from Agamben quickly, and I realize that the result may have been to allow an initial dissonance between Heidegger's and Agamben's notion of proper to pass unnoticed. Clearly Agamben does not follow Heidegger in an important respect, namely, Heidegger's location of improper on the side of one form of totalitarianism, namely, Leninism. Or better, Agamben draws out what was only suggested in Heidegger's analysis of Leninism in *Parmenides,* namely, that proper and improper are mutually inscribed in one another such that Western, historical man moves along an axis of a proper and improper relation to Being. Where the Soviet Union for Heidegger appeared to embody impropriety itself, Agamben adopts another perspective, or better, a second order of observation on proper and improper, so that what emerges is the process whereby the improper is excluded (be it the Jew excluded from the proper *bíos* of Germany or presumably the improper forms of collective life [petty bourgeois] in the West in what we might want to call, following his analysis of the baroque in *Stanzas,* "profane life"[9]). Yet it is unmistakable that in the association of improper with consumption and falsehood, Agamben nods to the associations that Heidegger draws between improper writing and its effects on making everyone the same.

Improper Means to a Coming Community

Another profound appropriation of the continuum between proper and improper appears in another of Agamben's texts. We can sense it in the first passage cited earlier, in which Agamben immediately identifies *zoē* as an improper form of life and then introduces another feature of *zoē* as encompassing its "sacralization." Here we note the parallel movement of *zoē* and the improper to extend their dominion as the latter begins to assume within it all other forms of life. In the following pages, Agamben will make passing reference to Bataille with regard to this making life improper, but those passages feel almost like an afterthought.[10] Instead, he rehearses that other move I noted previously when discussing improper writing and revelation in Heidegger, when Western man, now endangered by the withdrawal of Being thanks to these improper writing machines, is placed in the position of requiring saving—or in "The Question Concerning Technology," when Heidegger speaks of the danger of "Enframing" and in the same breath utters Hölderlin's verse about saving power. Heidegger offers few details about the form that this saving might take, but there is implicit in his association of the improper to saving an opening for introducing sacrifice into the dialectic of proper and improper. This is one of Agamben's most singular contributions to contemporary philosophy: the drawing forth of an implicit sacralization from Heidegger's ontology. That implicit sacralization will also become the basis for Agamben's thanatopolitical reading of industrial capitalism here and elsewhere as well as his unsettling reading of the camp as existing on both sides of the proper and improper divide.[11]

Much has been written about the notion of sacrifice in Agamben as it relates to the figure of *homo sacer*—a great deal of it turning on Agamben's further elaboration of the figure in *Homo sacer* and perhaps even more decisively in *Remnants of Auschwitz*.[12] The superimposition of *homo sacer* with the *Muselmann* in the later text has meant that when attempting to distill the features of Agamben's biopolitics, emphasis has naturally focused on the second half of the formulation I noted earlier: the process of making proper *bíos* ever more proper by excluding the improper, thus *zoē*. Agamben's own writings on the state of exception have reinforced such an emphasis.[13] The result has been to see Agamben's biopolitics as working primarily through a state of exception. This is, of course, true as far as it goes. But if we are to grasp Agamben's thanatopolitical perspective on contemporary life, the one in which impropriety is extended to all forms of life in the West, we need to look less to the state of exception and more to those texts in which technology takes center stage. Some of these pages will be found in *Homo Sacer* and *Remnants of Auschwitz*. Once done, I'll turn to Agamben's stunning recent work, in which the thanatopolitical features of his thought emerge even more clearly.

Before I do, a caveat: let's admit that it is not always so, that Agamben occasionally thinks technology in such a way that technology is not immediately transformed into a catastrophic power over life. A noteworthy example occurs in *The Coming Community* in the chapter titled "Without Classes." Here Agamben attempts to uncover the dialectic of proper and improper on a global scale, after the fall of the Berlin Wall and the emergence of what he calls "the planetary petty bourgeoisie."[14] After noting that contemporary man still lives in the shadow of Nazism, Agamben reintroduces once again the Heideggerian distinction between proper and improper relation to Being in ways that directly recall the Heidegger of *Parmenides* and his reading of "Heimkunft" in *Elucidations of Hölderlin's Poetry*:

> The petty bourgeois nullify all that exists with the same gesture in which they seem obstinately to adhere to it: They know only the improper and the inauthentic and even refuse the idea of a discourse that could be proper to them. That which constituted the truth and falsity of the peoples and

generations that have followed one another on the earth—differences of language, of dialect, of ways of life, of character, of custom, and even the physical particularities of each person—has lost any meaning for them and any capacity for expression and communication.[15]

Weaving strands of Heidegger's critique of technology here, Agamben replaces the Leninist metaphysician who inhabits a completely standardized technical world in which all are made the same with Western man. This "planetary petty bourgeoisie" has become so insensitive to all kinds of individual difference, Agamben will add, that what appears before the petty bourgeois seems all the same: difference has been flattened. The effect is similar to the anxiety that Heidegger felt in front of the typewriter but is now multiplied exponentially in Agamben's analysis. The result, again, is the domination of the improper over the proper, but here made much more intense. The references to language, dialect, and modes of living are to be expected, but what are not are "the physical particularities of each person," which now are no longer observed. We find ourselves deeply immersed in a terrifying world of technology, in which communication has biopolitical consequences in the sense that when communicating, individual difference is excluded, be it in speaking or, indeed, in how one looks physically.

At this moment, Heidegger will see two stars cross in the sky, one of ordering and the other of saving. Agamben, initially at least, differs little here from Heidegger. Speaking of death, he notes how "the petty bourgeois confront the ultimate expropriation, the ultimate frustration of individuality: life in all its nakedness, the pure incommunicable" and then raises the Heideggerian banner of death—the petty bourgeois is "probably the form in which humanity is moving towards its own destruction."[16] Thanks to communication, the petty bourgeois can now refuse any social identity, which is to say, are distanced from the mystery of Being (or of any homeland that could provide the coordinates for a return to Being). Only when "life in all its nakedness" comes to the fore, when "bare life" is utterly destined for destruction thanks to its improper relation to writing (or, if we prefer, to its proper relation to improper communication), does the saving power return. Now, however, it is thought not in terms of a catastrophic end of humanity but in terms of a future community in which one no longer speaks of an improper form of individuality. Rather one finds "a singularity without identity, a common and absolutely exposed singularity" or "singular exteriority."[17] It is no accident that Agamben's

The Coming Community remains his most persuasive piece on the future arrival of the common as it evades the thanatopolitics of so much more of his recent work.

Saving Power in *Homo sacer* and *Remnants of Auschwitz*

As exceptional as this moment is for marking Agamben's distance from the implicit thanatopolitics of Heidegger's ontology, Agamben more frequently declines Heidegger's saving power negatively as a Foucauldian power over life. We can see this most clearly in *Remnants of Auschwitz* and the overlap that Agamben's vision of thanatopolitics there has with his earlier volume *Homo sacer.* In *Remnants of Auschwitz,* Agamben sets this out in a series of readings of Primo Levi, Xavier Bichat, and Foucault. The most important passages concern what practically amounts to a minor genealogy of biotechnology thought through biopower and what Agamben sees as "the absolute separation of the living being and the speaking being, *zoē* and *bios,* the inhuman and the human."[18] Foucault's thinking of biopower and sovereign power launches Agamben's reflections:

> As we have seen, Foucault defines the difference between modern biopower and sovereign power of the old territorial State through the crossing of two symmetrical formulae. *To make die and to let live* summarizes the procedure of old sovereign power, which exerts itself above all as the right to kill; *to make live and to let die* is, instead, the insignia of biopower, which has as its primary objective to transform the care of life and the biological as such into the concern of State power. In the light of the preceding reflections, a third formula can be said to insinuate itself between the other two, a formula that defines the most specific trait of twentieth-century biopolitics: no longer either *to make die* or *to make live,* but *to make survive.*[19]

Leaving aside whether Agamben's reading of Foucault registers all the vagaries of the notion of biopower—he unmistakably elides biopower's inscription in the expansion of capitalism that occurred at the end of the eighteenth century—Agamben stakes out a position in which technology will be linked to the processes of subjectification and desubjectification that lead inexorably to a life that merely survives. The resulting "subject" is merely a shell that does not live in relation to others. Agamben gestures to this superimposition of Heideggerian ontology in a number of places, particularly when he inscribes possibility, impossibility, contingency, and necessity as "ontological operators," as "the devastating weapons used in the biopolitical struggle for Being, in which a decision is made each time on the human and the inhuman, on 'making live' or 'letting die.'"[20] This devastation recalls that other devastation of which Heidegger spoke in "Letter on Humanism": of language that extends and consumes ethical and moral responsibility, one that originates from a threat emanating from language itself. According to Heidegger, this threat to language also threatens existence and is based on its falling prey to "the service of expediting communication along routes where objectification—the uniform accessibility of everything to everyone—branches out and disregards all limits. In this way language comes under the dictatorship of the public realm which decides in advance what is intelligible and what must be rejected as unintelligible."[21] Differently from *The Coming Community*, Agamben, in *Remnants of Auschwitz*, translates Heidegger's profound worry about the effects of communication on Being into a political life *(bíos)* transcended by what is proper to a group and *zoē*, immanent to merely being alive.

The sheer importance of this distinction between *bíos* and *zoē* has, if anything, become more pronounced since September 11, 2001. As details of the nature of imprisonment and torture at Guantánamo and Abu Ghraib became known, what emerged were precisely the features of a biopolitical system whose intent clearly is to make a life that merely survives. And if and when Guantánamo is closed down permanently, it will in no small measure be due to the thought of Giorgio Agamben.[22] With that said, we risk losing the radicalism of Agamben's analysis if we simply slot it into a biopolitical critique of the war on terror. Said differently, we lessen the measure and import of Agamben's insight if we simply limit it to the biopolitical life (or lives) of terrorists. When Agamben speaks of a biopolitical struggle for Being, he evokes, consciously or not, threats to human existence that appear thanks to a change in man's relation to Being given the increasing role of technology. That Agamben moves through Foucault and his respective judgments on the relation between enunciation and existence shouldn't preclude us from noting that where Agamben winds up (or what he is left with) by the end of *Remnants of Auschwitz* is precisely the Heideggerian association of communication with improper writing. Thus the gloss of Foucault in which "enunciation is not a thing determined by real, definite properties; it is, rather, pure existence, the fact that a certain being—language—takes place" has the effect of reading Foucault's archaeology almost as if it were a Heideggerian instant of improper writing:[23] "Archaeology claims as its territory the pure taking place of these propositions and discourses, that is, the *outside* of language, the brute fact of its existence."[24] When Agamben speaks of existence or links a certain being to language, he has reenacted the originary division between proper and improper that structures Heidegger's ontology.

Yet where Heidegger continually draws the line between one kind of being who may be saved because Being has not sufficiently withdrawn to warrant the moniker Bolshevik or animal, Agamben marks that division within the subject herself; indeed, the modern subject who witnesses her own desubjectification is in fact what results from the superimposition of Western historical man with the Leninist metaphysician. Where enunciation and language are inscribed in "a certain being" that exists, the author merely looks on so as to recount the sheer breadth of her own ruin.[25] At this juncture, let's also note that when Agamben makes his final pitch for this reading of Foucault—that archaeology serves to point out the condition of separation in the subject between the *Muselmann,* "a living being," and the witness, "a speaking being"—he ends up with a radicalization of Heidegger's reading of proper and improper that now knows little internal limit to its ability to divide and separate life.[26] That Agamben will, at the end of *Remnants of Auschwitz,* rehearse Heidegger's saving power through a reading of remnant in Paul's "Letter to Romans" and then, later, in his *The Time That Remains* does not alter the impression that the notion of remnant is itself unable to cope with the onslaught of the improper against the proper, which language registers like some seismograph of twentieth-century biopolitical devastation.[27]

Agamben, of course, knows his debt to Heidegger's ontology based on proper and improper. Indeed, in the chapter "The Muselmann," Agamben, like Heidegger, also cites Hölderlin's principle "where there is danger, there grows the saving power," which he employs paradoxically in his reading of the camp as a site in which "every distinction between proper and improper, between possible and impossible, radically disappears."[28] He continues, "For here the principle according to which the sole content of the proper is the improper is exactly verified by its inversion, which has it that the sole content of the improper is the proper."[29] This elaboration of the improper sets apart the camps as well as Hitler's Germany as the moment when "biopolitics coincides immediately with thanatopolitics."[30] Agamben had earlier rehearsed the relation of proper and improper in *Means without End* when he discussed another set of proper and improper operators felt in the biopolitical division between people and population that consists "in transforming an essentially political body into an essentially biological body, whose birth and death, health and illness, must then be regulated."[31] The camp for Agamben will name this space in which the transformation of people into populations occurs and where populations are made over into *Muselmann.* Part of the reason the power of thanatos takes on so much weight in Agamben's reading is precisely because he superimposes a saving power—what in *Means without End* he associates with a Marxist society "without classes" or a "messianic Kingdom" in *Remnants of Auschwitz*—not simply over the camp itself but rather over all contemporary forms of life. We see this most clearly in *Means without End,* in those sections in which Agamben deconstructs the notion of people as a dialectical movement between two poles: "on the one hand, the *People* as a whole and as an integral body politic and, on the other hand, the *people* as a subset and as fragmentary multiplicity of needed and excluded bodies."[32] It is this fundamental division between bare life and political life that Agamben sees as structuring contemporary politics under the shadow of a modern thanatopolitics.[33]

Equally present, however, in this discussion of the camp extended to contemporary forms of life is Heidegger's notion of propriety and impropriety thought in relation to technology as the lever that, when pulled, allows Agamben's thanatopolitics to come into view. What Agamben has done is to have translated Being and being, proper and improper, handwriting and typewriting, into *bíos* and *zoē, Muselmann*/witness, and subjectification and desubjectification as the principal criteria for marking modernity. What gets lost in Agamben's translation of proper and improper from Heidegger into his own thanatopolitical perspective is, however, the role that technology plays in Being's withdrawal from man. In other words, in texts like *Means without End* and *Remnants of Auschwitz,* where the question is precisely the thanatopolitical valence of Nazism and, in particular, the camp, Agamben deanchors proper and improper from their moorings thought through a proper and improper form of writing. We see this in particular in *Remnants of Auschwitz,* in his recuperation of the poetic word as remnant, as "the one that is always situated in the position of remnant and that can, therefore, bear witness," which implies that Agamben has not only essentially affixed to the proper the poetic word but has translated and updated it as bearing witness.[34]

The question that we will want to raise is this: what happens to improper forms of writing when the proper has essentially been diluted so as to be able to include within it the possibility of testimony? Agamben would surely argue that if the poetic word can still function as a vehicle for a remnant, it is because it is still possible for a proper form of writing to coordinate movement from man to Being; or rather, in a moment when the proper is put everywhere at risk by the improper, a proper form of writing, the poetic word, can now only bear witness to the domination of the improper over the proper. The uncomfortable conclusion that emerges from this reading of Agamben would be its deep indebtedness to a Heideggerian ontology of proper and improper writing precisely when the testimony concerns the camp itself. The effect is once again to extend the biopolitical devastation across modernity so that the form of proper writing today, testimony, is only a specter of its earlier "self." In other words, testimony, in some sense, becomes the privileged mode of a new genre to be called *thanatopolitical writing.*

What has not changed with regard to Heidegger's understanding of the function of the poetic word, however, is the context of proximity, which is to say, not merely to bring the reader of testimony nearer the event of desubjectification but to allow the subject herself to take up the necessary distance before the *Muselmann*'s (or her own) elaboration of desubjectification. What is required for testimony will be the continual movement between nearness and distance that proximity names as a continuum, a proximity offered by dead language: "If we now return to testimony, we may say that to bear witness is to place oneself in one's own language in the position of those who have lost it, to establish oneself in a living language as if it were dead, or in a dead language as if it were living."[35] Yet proximity for Agamben is little different from Heidegger in the sense that "proximity to the source is mystery," as Heidegger notes in his essay on Hölderlin. For Agamben, this proximity to the desubjectification of the subject who bears testimony to it is also mysterious: it is the mystery in which the speech of the author and the witness bear "witness to a time in which human beings did not yet speak; and so the testimony of human beings attests to a time in which they were not yet human."[36]

That Agamben has recently backtracked from some of these claims—in making modernity coterminus with the production of the *Muselmann* and *homo sacer*, for instance—becomes apparent in a recent essay titled "What Is a Paradigm?" collected in *The Signature of All Things* and which, given its title, is surely intended as a bookend to his other recent essay *What Is an Apparatus?* Here Agamben attempts to elude his critics, especially those who find his reading of thanatopolitics, the camp, and our modernity as too inclusive of death. Without lingering too long over the text, I do want to note that Agamben's defense of his work as evolving through a series of paradigms not only draws on Heidegger for its authorization but indeed inscribes the very notion of paradigm in ontology: "If one asks whether the paradigmatic character lies in things themselves or in the mind of the inquirer, my response must be that the question itself makes no sense. The intelligibility in question in the paradigm has an ontological character. It refers not to the cognitive relation between a subject and object but to being."[37] And indeed, a page before, Agamben notes that "*Homo sacer* and the concentration camp, the *Muselmann* and the state of exception . . . are not hypotheses through which I intended to explain modernity by tracing it back to something like a cause or historical origin."[38] Still Agamben cannot have it both ways, which is to say that, these statements to the contrary, as long as the reference to ontology and being continues to signify how a paradigm makes sense, at the heart of that ontology is the difference between proper and improper that, when not explaining modernity in the sense of cause and effect, does shift the terms to some earlier transhistorical framework not so distant from Heidegger's own critique of technology.

Homo sacer Across the Ages

To sum up, Agamben's thanatopolitical reading of contemporary biopolitics moves through an appropriation of Heidegger's categories of proper and improper writing. This occurs, on one hand, through the reading of testimony as a spectral form of proper writing in which the desubjectification of the subject writes down the steps through which his being has been devastated. Agamben has translated the deprivation of the hand enacted by mechanical writing onto the subject, but where, for Heidegger, the character of the subject of proper writing was concealed by the typewriter, with oblivion of Being as its result, for Agamben, proper writing in the biopolitical space of the camps is not understood as some separate form of writing protected from the standardization of communication but rather as deeply embedded in the realm of the improper. For that reason, proper writing principally testifies to its own impossibility or its own degraded "essence." As a result, the poetic word, that earlier bastion of proper writing that Heidegger discussed, is seen as profoundly fractured by a twentieth-century thanatopolitics. It becomes a mere remnant of itself. On the other hand, there is in Agamben's perspective another vector that moves simultaneously through the body politic of our own contemporaneity and that concerns the notion of standing reserve *(Bestand)*, a term Heidegger puts forward in "The Question Concerning Technology," as well as the mutual inscription of distance and proximity that is at the heart of Heidegger's reading of Hölderlin. This second perspective on technology, proper writing, and proximity has emerged more clearly of late in an essay of Agamben's as well as in his latest volume of *Homo sacer*, titled *The Kingdom and the Glory*.

Many have written on Agamben's use of the term *homo sacer*, from his appropriation of the term from ancient Rome and the laws governing the Roman *pater* with respect to his sons to his overlaying of *homo sacer* with contemporary man today.[39] The question I want to ask concerns more the outlines of thanatopolitics when Agamben's object is contemporary Europe. To do so, I want to focus less on Agamben's analysis of the camp as "an event that marks in a decisive way the political space itself of modernity," which undoubtedly has the effect of collapsing the political space between both totalitarianisms of the left and right (as well as liberal democracies and totalitarianisms), and instead examine those more avowedly thanatopolitical moments in which technology reemerges as central.[40] These moments of thanatopolitics in Agamben's more recent writings do not fall neatly into the extension of the space of the camp to our own contemporaneity when, for instance, Agamben writes, again in *Means Without End*, that "we can expect not only new camps but also always new more delirious normative definitions of the inscription of life in the city. The camp, which is now firmly settled inside it, is the new biopolitical *nomos* of the planet,"[41] or when Agamben himself seems to authorize the kind of pressure of a term like *homo sacer* in *Homo sacer*, when he extends it to all of us: "Sacredness is a line of flight still present in contemporary politics, a line that is as such moving into zones increasingly vast and dark, to the point of ultimately coinciding with the biological life itself of citizens. If today there is no longer any one clear figure of the sacred man, it is perhaps because we are all virtually *homines sacri*."[42] Where "a life that as such is exposed to violence without precedent in the most profane and banal ways"—the example Agamben ironically offers is of weekend deaths in automobile accidents—then it would indeed appear that all of us are living globally in a camp to the degree we can be killed with impunity as *homo sacer*.[43]

Many of Agamben's critics have been troubled by this extension of *homo sacer* to contemporary man. To my mind, it is one of the principal reasons explaining why Agamben's subtitle of the third part of *Homo sacer,* namely, "The Camp as Biopolitical Paradigm of the Modern," has effectively limited Agamben's critique of modernity to a forced superimposition of the camp and the city, of contemporary man and *homo sacer* and *homo sacer* and *Muselmann.* Therefore let's also recall that Agamben repeatedly reserves the term *Muselmann* only for the Nazi camps and not for present-day *homo sacer. Muselmann* would thus appear to name a subgenus of *homo sacer.* When seen against the extension of the camp, a tension results with regard to the object of Agamben's critique of *homo sacer.* Or, formatted as a question, might there be lurking in the crushing of distinctions between our contemporaneity and Nazism another perspective on the thanatopolitics of liberalism (and not Nazism), say, that would not or could not be immediately inscribed in the horizon of the camp—that would move outside the biopolitical paradigm that Agamben employs in *Homo sacer,* for instance?

Disposing of Life

Agamben adopts such a perspective in his most recent work, *The Kingdom and the Glory,* as well as in a shorter text that appeared in conjunction with it, titled *What Is an Apparatus?* I'd like to focus on the latter text initially because it abbreviates the most important thanatopolitical moments of Agamben's recent thought in a form that resembles both a manifesto (of methodology, we should be clear) and a summing up of Agamben's own use of the term (especially when we recall that it was Deleuze who, almost twenty years earlier, had written a tract with the same title). I want to read this text with two questions in mind. First, how is the notion of *dispositif* to be thought in relation to *homo sacer,* or better, how are we to think what Agamben calls the theological *oikonomia* with a government that is intent everywhere on augmenting the number of *dispositifs* at its disposal, not merely to govern man, or to save him from catastrophe, but rather to expedite the catastrophe itself of contemporary man's essentially profane nature? Second, how are we to think this implicit thanatopolitical valence of the *dispositif* through the terms of the first chapter, which is to ask, what role does Heidegger's "tragic" ontology, as Negri calls it, play in accounting for the thanatopolitics of contemporary liberal and neoliberal thought? The answer, obviously, is a great deal, which explains the sheer power of Agamben's essay in laying bare the modes by which liberalism, through a theologically inflected notion of economy, creates and continues to create subjects subject to desubjectification, thanks to the proliferation of *dispositifs*.

I noted earlier that Deleuze also wrote a small tract asking what is a *dispositif*, and although Agamben's perspective follows a different itinerary than Deleuze's, a brief comparison between them will help get things under way. The initial framework for both Deleuze and Agamben concerns the use that Foucault makes of *dispositif* and its relation to the production of subjectivity. For both, the launching point are those pages Foucault dedicates to it in *Dits et écrits*:

> By the term "apparatus" I mean a kind of formation, so to speak, that at a given historical moment has as its major function the response to an urgency. The apparatus therefore has a dominant strategic function. . . . I said that the nature of an apparatus is essentially strategic, which means that we are speaking about a certain manipulation of relations of force, of a rational and concrete intervention in the relation of forces, either so as to develop them in a particular direction, or to block them, stabilize them, and to utilize them. The apparatus is thus always inscribed into a play of power, but it is also always linked to certain limits of knowledge that arise from it and, to an equal degree, condition it. The apparatus is precisely this: a set of strategies of the relations of forces supporting, and supported by, certain types of knowledge.[44]

For Deleuze, the *dispositif* cannot be thought apart from lines of force or flight that condition the process of subjectification. Thus "Foucault, for his part, was concerned that the social apparatus [*dispositifs*] which he was analyzing should not be circumscribed by an enveloping line, unless other vectors could be seen as passing above or below it."[45] The result of these lines of flight is a mobile notion of self that shifts as the process of subjectification shifts:

> This dimension of the Self is by no means a pre-existing determination which one finds ready-made. Here again, a line of subjectification is a process, a production of subjectivity in a social apparatus [*dispositif*]: it has to be made, inasmuch as the apparatus allows it to come into being or makes it possible. It is a line of escape.[46]

Deleuze's understanding of *dispositif* will allow him to move outside the fixed contours of a ready-made Self (and subject) and to focus on "a process of individuation which bears on groups and on people, and is subtracted from the power relations which are established as constituting forms of knowledge."[47] I want to return to this Deleuzian reading of *dispositif* and individuation shortly when discussing Roberto Esposito's use of the term, but for now, we can already note a major difference between Deleuze's and Agamben's reading of *dispositif.* Nothing initially in Deleuze's account evokes implicitly the negative. The *dispositif* conditions the production of subjectivity but also highlights the lines along which the produced subjectification creates lines of flight that will in turn come together in other *dispositifs.*

Whereas Deleuze suggests (though does not name) the possibility of an affirmative biopolitics with individuation by focusing on those moments in Foucault's thought, principally *Hermeneutics of the Subject,* in which the city becomes a *dispositif* for subjectification, Agamben instead will draw on a different Foucault for his understanding of *dispositif,* one based on a separation between subjects that makes government possible. Here I have in mind an interview with Foucault from 1974 titled "The Risks of Security." Speaking of postindustrial societies and welfare, Foucault notes, in terms that directly recall Agamben's own, the importance of separation for security:

> No doubt we can say that certain phenomena of marginalization are linked to factors of separation between an "insured" population and an "exposed" population. Moreover, this sort of cleavage was foreseen explicitly by a number of economists during the seventies, who thought that in postindustrial societies the exposed sector would, on the whole, have to grow considerably. . . . There are in certain forms of marginalization what I would call another aspect of the phenomenon of dependency. Our systems of social coverage impose a determined way of life that subjugates [*assujettit*] individuals. As a result, all persons or groups, who for one reason or another, cannot or do not want to accede to this way of life themselves are marginalized by the very game of the institutions.[48]

The passage stands out for its echoes with *zoē* and *bíos,* the marginalized and exposed who "for one reason or another" remain outside a particular "way of life." It seems, then, that Agamben has extended backward the moment of social coverage that Foucault describes to some unknowable and unlocalizable point in the past such that governing always involves separation. Foucault authorizes this interpretation with his use of the term *population,* which, for Agamben, suggests the contemporary biopolitical specter of "people." But more is at work here than that. Indeed, Agamben's reading of *dispositif* ought to be thought of as a search for "factors of separation" across modernity, beginning (and ending) with political theology. In *The Kingdom and the Glory* and *What Is an Apparatus?,* Agamben will use this reciprocity between separation and *dispositif* as the basis for his thanatopolitical reading of liberal, industrial democracies.[49] In these democracies, technology most profoundly alters the very meaning of separation and the common, or to use Agamben's more current dialectic, technology is seen as integral to a governing that utterly depends on a conception of the profane and sacred. For Agamben, the profane and the sacred powers govern through the medium of the *dispositif.* He writes,

> From this perspective, one can define religion as that which removes things, places, animals, or people from common use and transports them to a separate sphere. Not only is there no religion without separation, but every separation contains or conserves in itself a genuinely religious nucleus. The apparatus [*dispositif*] that activates and regulates separation is sacrifice. Through a series of minute rituals that vary from culture to culture (which Henri Hubert and Marcel Mauss have patiently uncovered), sacrifice always sanctions the passage of something from the profane to the sacred, from the human sphere to the divine. But what has been ritually separated can also be restored to the profane sphere. Profanation is the counter-apparatus that restores to common use what sacrifice had separated and divided.[50]

Note the emphasis Agamben places on *dispositif* and separation, which are presumed in the very nature of religion. Agamben assumes a form of religion that is coterminus not with the sacred but with determining both the sacred and profane (because one cannot be decided without implicitly naming another). Here, as Agamben notes, he is drawing on the work of Marcel Mauss, as he did previously, and so it is not surprising that when speaking of religion, it becomes the means for the withdrawal of places, animals, and persons from a "common use." Indeed, in *Means without End,* Agamben had even posed the question about the nature of "common use" in industrial democracies today—"How does one use a common?"—and the answer in *What Is an Apparatus?* is that one uses it only after it has been separated from the profane, which is now homologous with the common, or with "common use."[51] Paradoxically, then, where we might have expected to see community in league with religion, especially given the Paulist tradition that so interests Agamben, instead Agamben sees religion on the side of separation; indeed, any form of separation is inscribed in the horizon of religion.[52] Agamben will devote much more space and detail to these reflections in *The Kingdom and the Glory,* but meriting our attention is how he arms religion with *dispositifs,* the chief of which is sacrifice deployed across rituals. It is in these rites that the move from the profane to the sacred occurs—different words for creating separation and proximity (the sacred and profane) by which things, places, animals, and subjects are governed.

Agamben draws two inferences from this association of religion (and *dispositif*) with separation. First, echoing again Deleuze's reading of *dispositif,* Agamben will expand the limits of what he means by *dispositif* well beyond the one limited to religion. Thus he posits that "every apparatus implies a process of subjectification, without which it cannot function as an apparatus of governance, but is rather reduced to a mere exercise of violence." Soon after, he adds that "apparatus, then, is first of all a machine that produces subjectifications, and only as such is it also a machine of governance."[53] Note how Agamben sets up a relation between subjectification and *dispositif,* as Foucault and Deleuze do, but inflects it toward the sacred. Here subjectification is viewed as succeeding (or better, as superimposed over) the separation and making sacred of the religious *dispositif.* The process of subjectification and desubjectification cannot be thought apart from the category of the *dispositif,* whose origins reside precisely in the difference religion posits between the profane and the sacred. All this echoes much of Agamben's previous thought on *homo sacer,* while appearing to overlay the functioning of the *dispositif* over the anthropogenetic machine that creates humans out of animals and vice versa.

Yet consider as well how Agamben connects *dispositif,* its role in subjectification, and its concomitant function as a machine of governing. The appearance of the term *government* returns our attention once again to Foucault, as Agamben himself notes in the opening to the pamphlet, but with a twist. Whereas government for Foucault referred both to a governing of oneself in *Hermeneutics of the Subject* and of the state (hence the art of governing that would cover both), Agamben links *dispositif* and government to another moment not so easily localizable in European history (as Foucault does both for pastoral power, which characterized the seventeenth and part of the eighteenth century, and for biopolitics, which emerged at the end of the eighteenth century).[54] Agamben suggests a religious, indeed, a sacred inflection to the notion of governing to the degree governing cannot be thought apart from the proliferation of *dispositifs.* He makes two moves, then: on one hand, an almost universalizing notion of *dispositif* that comes to us from the religious distinction between profane and sacred, and on the other, an implicit marking of contemporary forms of governing as involving the multiplication of *dispositifs* that allow for the movement of a host of beings and things from the sphere of the profane to the sphere of the sacred.

Liberal Thanatopolitics

Before turning to the thanatopolitical effects of this reading of *dispositif* and governing, I want to complicate this reading in ways Agamben does not by introducing another reading Foucault gives of governing in *The Birth of Biopolitics* when the subject is liberalism. Writing about the emergence of a liberal form of government, Foucault speaks of "a new art of government" whose essential characteristic is "the organization of numerous and complex internal mechanisms whose function . . . is not so much to ensure the growth of the state's forces, wealth, and strength, to ensure its unlimited growth, as to limit the exercise of government power internally."[55] I don't see anything incompatible between Agamben's perspective about governing and separation, or governing as separating the sacred from the profane, from what Foucault argues earlier. I do want to suggest, however, by way of comparison, that here, at least in Foucault's most extensive statement on liberalism, we don't find the thanatopolitical inflection that one finds full-blown in Agamben—that, in fact, in *The Birth of Biopolitics,* Foucault draws clear limits around any resemblances between Nazism, for instance, and German liberalism after the war. This is proof, I think, that for Agamben, unlike Foucault, governing implicitly works through *dispositifs,* whose primary effect is to mark biopolitical divisions between fully formed subjects and their spectral de-subjects. The *dispositif* for Agamben is a machine that produces subjects so that it is becoming increasingly difficult to locate where government ends and *dispositif* begins. Thus the result at the end of *What Is an Apparatus?* is precisely the "ungovernable."[56]

What, then, characterizes the differences between *dispositifs* today and their long history of furthering the religious work of *oikonomia*? To answer this question allows us to see precisely where Agamben updates (if that is the correct word) Foucault's reading of governmentality and liberalism by emphasizing the *dispositif* as an instrument of proper and improper writing and revelation. He does this repeatedly, but most spectacularly when he argues that something has radically changed in the nature of the *dispositif* in the current phase of capitalism such that it is no longer subjectivity but rather its opposite, desubjectification, that is produced. The process through which this occurs is one we do well to dwell on:

> What defines the apparatuses that we have to deal with in the current phase of capitalism is that they no longer act as much through the production of a subject, as through the processes of what can be called desubjectification. A desubjectifying moment is certainly implicit in every process of subjectification. As we have seen, the penitential self is constituted only through its own negation. But what we are now witnessing is that processes of subjectification and processes of desubjectification seem to become reciprocally indifferent, and so they do not give rise to the recomposition of a new subject, except in larval or, as it were, spectral form.[57]

Agamben believes that the spectral nature of the subject today emerges out of the self-canceling movement between subjectification and desubjectification that is produced thanks to the proliferation of *dispositifs* that the current stage of capitalism creates (though as the reader can judge for herself, much of Agamben's critique here is not really capitalism per se but rather the multiplication of *dispositifs* in ways unheard of in a history of providential government). The argument that subjectification carries out desubjectification is one we have come to expect from Agamben. As we saw, the implicit condition for the emergence of the two figures of *Muselmann* and he who offers testimony was desubjectification; it was also the basis for the sense of messianism that characterizes Agamben's notion of remnant. But here Agamben's point of reference is no longer officially Nazism and its attempts to rid the body politic of the improper, nor is it industrial democracy, which expands the improper at the expense of the properly political, though there are clearly echoes of this when Agamben writes that the primary effect of governing through *dispositifs* today is the eclipse of politics. Agamben will speak of capitalism and modernity as well as contemporary society, but lurking all the while is a notion of *oikonomia*. What explains this change in the nature of the *dispositif* is its explosive proliferation. Two effects result. First, the aforementioned eclipse of the political à la Arendt that we know so well, but prior or even contemporary with it, *dispositifs* which before produced subjects now produce desubjects. What is beyond doubt for Agamben is that under the current regime of *dispositifs* that expand uncontrollably, the properly negative (and not, we note, the improperly negative) turns on itself, thus leading to the emptying of the subject. Whether this is implicit in the notion of subject itself at its originary moment of formulation—for Agamben, it is given the deconstruction that he offers of proper and improper, following Heidegger in *Means without End*—what matters is that under the explosion of *dispositifs,* the subject is turned inside out like a glove, the result being the larval, spectral form of (de) subjectified subjects as well as a decisive move away from a healthy body politic to "the most docile and cowardly social body that has ever existed in human history."[58]

Resisting the *Dispositifs* of the Thanatopolitical

How extensive is this proliferation of *dispositifs*? We are far beyond Foucault's panopticon, it would seem: "We could say that today there is not even a single instant in which the life of individuals is not modeled, contaminated, or controlled by some apparatus."[59] Thus, for Agamben, *dispositif* will literally name anything that has "in some way the capacity to capture, orient, determine, intercept, model, control, and secure the gestures, behaviors, opinions or discourses of living beings."[60] We are no longer simply dealing with the naive intersubjectivity of contemporary social philosophy that emphasizes recognition protocols that have come down to us from Hegel as the means by which subjectivity is born. We sense that in the slight shock that arrives as we read "living beings." Nor are we dealing with subjects or desubjectified subjects but rather with living beings who suffer the misfortune of having been captured. Agamben provides the example of the primate, blissfully unaware of the consequences, that "inadvertently let himself be captured, probably without realizing the consequences that he was about to face."[61] Thus, from Agamben's perspective, there are no subjects without *dispositifs,* but equally no desubjectified subjects without *dispositifs.*

Here we have an important resemblance between Agamben's biopolitical perspective on the *dispositif* and Sloterdijk's use of the term *media* (domesticating and bestializing), which I will be discussing in the following chapter. The assumption for both is a stratum of life on which *dispositifs* act improperly and impolitically. Still, a significant difference can be found between Sloterdijk and Agamben here. Agamben does not argue, as Sloterdijk will in "Rules for the Human Zoo," that the properly human results only from domesticating media, or translated into the terms of the *dispositif,* Agamben does not limit subject formation merely to the encounter between *dispositifs* and living beings. He does find a place for subjectivity to emerge, as well, from the relation between bodies: "To recapitulate, we have then two great classes: living beings (or substances) and apparatuses. And between these two, as a third class, subjects. I call a subject that which results from the relation and, so to speak, from the relentless fight between living beings and apparatuses."[62] Although Agamben does not offer any further details on the interaction "from body to body" that, in conjunction with *dispositifs,* produces subjects, on the next page, he elides the contact between bodies when he poses the obvious question about the best form of resistance in a quotidian corporeal battle with *dispositifs.* It would be helpful to hear more about this implicit space and contact between living beings that precedes and so escapes the introduction of the *dispositif.* My own questions turn on the interval of separation and also contact between living beings and how one might try to speak about this space, let alone fortify it. I'll have more to say later in this chapter, when I turn to Roberto Esposito's use of *dispositif* and, what's more, the idolatrous critique he offers of Agamben's interpretation of *dispositif.*

It is true that Agamben will go on in the essay, and at much greater length, in *The Kingdom and the Glory* as well as in *Profanations,* about how best to resist the exponential multiplication of *dispositifs* today. What is required is a counterapparatus that is capable of restoring "to common use what sacrifice had separated and divided."[63] This would presumably work by breaking out of the division between *proprium* and *improprium*—by distinguishing between what is properly one's own (be it ethnic group, state, or individual property owner) and what is not. For Agamben, we recall that this becomes possible in *Means without End,* however, only by refashioning our understanding of the common as "a point of indifference between the proper and the improper—that is, as something that can never be grasped in terms of either expropriation or appropriation but that can be grasped, rather, only as *use.*"[64] As important as these considerations are for thinking beyond ever increasing separation, let's linger over those moments in which current *dispositifs* dominate being through separation because it is there that we'll find Agamben's Heideggerian-induced perspective on the thanatopolitical.

The thanatopolitical arises in Agamben's discussion at precisely the moment when the *dispositif* separates. This act of separation operates on two registers: in the domain of the theological and then in the separation within man between the human and the animal. In the first, the *dispositif* in its theological configuration creates a division between being and praxis, "the nature or essence, on the one hand, and the operation through which He administers and governs the created world, on the other."[65] God himself, in this view, is only able to govern the world of the living to the degree he contains within himself a division between being and praxis. Being alone is not sufficient to allow God to administer or govern because apparently one does not govern being but rather actions alone. Thus, of God and being, Agamben writes, "Action (economic, but also political) does not have any foundation in being; this is the schizophrenia that the theological doctrine of *oikonomia* leaves as its inheritance to Western culture."[66] With the emphasis on action, we'll certainly want to recall those pages in *Parmenides* I discussed earlier, in which Heidegger uncovers the hidden relation among hand, action, and activity. The key moment concerned the features of action, *Handlung* in German:

> This word [*Handlung*], however, does not mean human activity *(actio)* but the unitary way that at any time things are on hand and at hand, i.e., are related to the hand, and that man, in his comportment, i.e., in his acting by means of the hand, is posited in relation to the things.[67]

What Agamben has done is to have made homologous the workings of the hand with the *dispositif* such that where the hand is what "secures the reciprocal relation between 'beings' and man," the *dispositif* does so between living beings and subjects.

The second moment of separation occurs when Agamben rereads Heidegger's concept of the open through the lens of *oikonomia. Dispositifs,* Agamben will argue, are intimately linked to the process of "humanization that made 'humans' out of the animals we classify under the rubric Homo sapiens."[68] Here Agamben moves in an unexpected direction, finding the distinction between being and action that structured theological governance operating, as well, in the coming into being of man. "The event that produced the human constitutes, for the living being, something like a division, which reproduces in some way the division that *oikonomia* introduced in God between being and action."[69] Thus the living being is now separated from the animal who acts to create a world, in a process that Agamben describes at length in *The Open.* There the human emerges from the animal by breaking the interaction with the disinhibitors that goes under the name of "boredom." Here the series *oikonomia,* governing, and *dispositif* is superimposed over the underlying break in which the human is constructed and distinguished from a mere living being, and it is seen as part of the same process underlying theological forms of governance.

Dividing Life, Drifting to Death

Much more remains to be said about this process of separation and its effects on the overall thanatopolitical tenor of Agamben's thought. And I'll want to compare it in due course to Peter Sloterdijk's discussion of "Clearing" in "Rules for the Human Zoo" and to "Domestikation des Seins," for their mutual indebtedness to Heidegger is clear enough. For now, let's consider the description Agamben offers of the working of the *dispositif* in a context of governability and ungovernability that resonates deeply with what I see as a problematic drifting toward the thanatopolitical in Heidegger. Now Agamben, of course, readily admits the Heideggerian inflection of his use of the term *dispositif.* In one of two extended paragraphs dedicated to Heidegger, Agamben notes how the etymology of the term *Gestell* is "similar from an etymological point of view to *dispositio, dis-ponere,* just as the German *stellen* corresponds to the Latin *ponere.*"[70] He will hear it, too, in *Bestellen,* or the "ordering" that informs so much of Heidegger's reading of technology. The result is *oikonomia:* economy crossed with theology, Foucault, and Heidegger's *Gestellen–Bestellen* that will encompass all those knowledges and practices that govern the behavior and thoughts of man. Yet more is at play here than that. In the previous chapter, I noted how according to Heidegger, in a regime of modern technology, Being is withdrawn because of the effects set in motion by improper writing, one of the most important of which is to turn man into "one ordered" by technology. So far, we remain in the realm of *Gestellen,* as Agamben himself notes. But Heidegger will also speak of another term related to *Gestellen–Bestellen,* which will account for technology's capacity to keep man in place. Writing in "The Question Concerning Technology," Heidegger reminds us that

> the word *stellen* [to set upon] in the name of *Ge-stell* [Enframing] not only means challenging. At the same time it should preserve the suggestion of another *Stellen* from which it stems, namely, that producing and presenting [*Her- und Dar-stellen*] which, in the sense of *poiēsis,* lets what presences

> come forth into unconcealment. This producing that brings forth—e.g., the erecting of a new statue in the temple precinct—and the challenging ordering [*Bestellen*] now under consideration are indeed fundamentally different, and yet they remain related in their essence. Both are ways of revealing, of *alētheia.* In Enframing, that unconcealment comes to pass in conformity with which the work of modern technology reveals the real as standing-reserve [*Bestand*].[71]

The overall scheme of *Gestellen* that separates man from Being, the figure that emerges of standing reserve, is held in place through a process of an improper revelation, an improper *Herstellen* and *Darstellen. Gestellen* contains both within it as the means by which man is held in place or captured, as Agamben translates it. We recall that it was this being on hold or at hand for technology (and so to become the subject of an unending loop of improper revelations) that spelled trouble for Heidegger because it instituted new and improper relations everywhere between man and an increasingly remote Being. Humanity is made to be on call as a standing reserve, made to wait for a series of revelations whose only real consequence is to place man in the position of those who are ordered.

The drift toward the thanatopolitical that we saw taking place in Heidegger as a result of the effects of improper writing on Being is, not surprisingly, therefore, to be found here in Agamben as well, and given the explosion of *dispositifs* in his reading of contemporary society, we can expect the catastrophe to be even more shattering than the one noted but never directly recounted by Heidegger. And here Agamben does not disappoint. The transformation of Heidegger's critique of technology into the perspective of the *dispositif* now moves well beyond Foucault's interpretation of the *dispositif* to include almost every knowledge, practice, measure, and institution that makes useful "the behaviors, gestures, and thoughts of human beings."[72] Agamben himself acknowledges the catastrophe that awaits us in a regime of never-ending *dispositifs* and the governmental machinery that can no longer control them: "Rather than the proclaimed end of history, we are, in fact, witnessing the incessant though aimless motion of this machine, which, in a sort of colossal parody of theological *oikonomia,* has assumed the legacy of the providential governance of the world; yet instead of redeeming our world, this machine . . . is leading us to catastrophe."[73] The catastrophe is measured not in the sheer numbers of *homines sacri*—in fact, the references to *homo sacer* are completely missing in the essay—but rather in the biopolitical effects on the body politic. Thus "contemporary societies . . . present themselves as inert bodies through massive processes of desubjectification without acknowledging any real subjectification."[74] The catastrophe for Agamben is the production of millions of inert bodies that have forgotten, in Heideggerian terms, the essence of proper action and who thus are in need of saving but who are led instead by the contemporary form of *oikonomia,* that is, governmental machinery in the guise of the providential machine, "to catastrophe."[75]

The catastrophe is twofold. First, these desubjectified subjects may be killed or abandoned in greater numbers than ever before. Second, attempts by governmental machinery to administer life that has already been made docile, depoliticized, and desubjectified are, in a word, destined to fail. It is as if the proliferation of *dispositifs* has shoved a stick in the wheel of contemporary schemes of governing such that rather than administering life, government essentially administers death to its former citizens. Although Agamben does not offer any but the most passing of details, his anxiety about biometric scans and fingerprinting evokes the thanatopolitical specter of Nazism.[76] The catastrophe is that in attempts to administer life today, the *dispositifs* that make that possible have so thoroughly made lives docile that governability itself is placed in doubt. When that occurs, catastrophe is only a matter of time in the time that remains to us as remnant.

A Catastrophe Foretold

With more time, we surely would want to juxtapose Agamben's catastrophe of *dispositifs* gone wild in a current configuration of the thanatopolitical that goes under the name of *oikonomia* with Antonio Negri and Michael Hardt's reading of empire and, concurrently, the multitude. Agamben implicitly poses the impolitical question of how it is possible to speak of a multitude that enjoys astounding reserves of biopower when the effect of the proliferation of *dispositifs* is to create prone bodies who merely obey and who will willingly allow themselves to be scanned, fingerprinted, and watched so as to make their own sacrifice all the easier. I return to some of these questions in the following chapter, but in the meantime, and as a way of bridging to the final part of my discussion on the thanatopolitical and contemporary Italian thought in the work of Roberto Esposito, I do want to add something else about the power of the thanatopolitical in Agamben's more recent work. It is important to observe in this regard that Heidegger, too, made technology a fundamental operator in his ontology, but in both *Parmenides* and "The Question Concerning Technology," he continues to privilege a certain form of technology that goes under the name of improper writing and communication. By extending technology to include all those *dispositifs* that create docile bodies, Agamben risks being unable to distinguish the specific features of communication technologies that endanger man, in particular, the power of technology in the form of communication to make individuals the same so as to be better able to slot them into points in a communication network. What are the benefits of collapsing the cellular phone, for instance, into a larger and undifferentiated category of the *dispositif* that includes contemporary gadgets as well as the accessories of the first *Homo sapiens*? The effect in making it impossible to see where the *dispositifs* of antiquity end and modern ones begin is to dehistoricize to such a degree that the human is always already captured by the *dispositif*—that the human from its origin is already destined for salvation, which, for Agamben, can only mean catastrophe. Recall those pages of *The Open* in which Agamben speaks of the political tasks that still retain some seriousness today. Faced with this eclipse, "the only task . . . is the assumption of the burden—and the "total management"—of biological life. Genome, global economy, and humanitarian ideology are the three united faces of this process, in which post-historical humanity seems to take on its own physiology as its last, impolitical mandate. . . . The total humanization of the animal coincides with a total animalization of man.[77]

In *The Kingdom and the Glory* and *What Is an Apparatus?*, the management of biological life will be translated as the modern equivalent of *oikonomia*, but as I suggested earlier, one can easily extend this reading of the current "total management of biological life" to that other twentieth-century thanatopolitical management style that was Nazism. Nazism, of course, attempted to harness the biopower of Germany, as Foucault noted in "*Society Must Be Defended*," in ways that built on nineteenth-century attempts by power to take life under its care. It did so by crossing biopower with sovereign power:

> The two mechanisms—the classic, archaic mechanism that gave the State the right of life and death over its citizens, and the new mechanism organized around discipline and regulation, or in other words, the new

mechanism of biopower—coincide exactly. We can therefore say this: The Nazi State makes the field of life it manages, protects, guarantees, and cultivates in biological terms absolutely coextensive with the sovereign right to kill anyone, meaning not only other people, but its own people.[78]

Total management of life under Nazism was possible not only because of the new mechanism of biopower. What transformed biopower into thanatopower and the practices of thanatopolitics was the reeruption of sovereignty, of both being deployed over the field of life. Furthermore, for Foucault, management of life could only take place when not just death (which is everywhere implicit in the terms of biopower) fell under the Nazi State's purview but when one's own people could be murdered, which is to say, when those considered as a people proper could now be killed. In other words, the biopolitics of treating one's own people improperly (as a population and not as *bíos*) did not provide the State with the means to manage the field of life as extensively as it wanted. Only the sovereign right to kill provided the means to practice a thanatopolitics over the proper.

How do Agamben's reflections take up these considerations of the thanatopolitical? For Agamben, for a total management of biological life to be possible, biopolitics is not enough. What is required is the deployment of sovereignty over life and death. One cannot manage life without managing death—this is for Agamben Foucault's principal lesson—and managing death does not require simply care for life and reproduction but also employing death. And as Agamben himself notes, no power was more intimately linked to maintaining the proper, to a proper body politic or people, than sovereignty. If there is a catastrophe foretold, it concerns the possibility that sovereignty today will somehow endanger those inert and docile bodies created by *dispositifs* who no longer wish or are no longer able to resist. Agamben confirms this reading in *What Is an Apparatus?* when he never actually makes the proliferation of *dispositifs*—whose effect is to make relations between persons more abstract—the *means* by which the catastrophe will take place. For that to happen—and here I'm pushing Agamben's argument where it seems to me implicitly to be heading—a sovereign power to take life is always ready to be unleashed at the precise moment when the extension of inert bodies has reached such numbers that sovereignty can do nothing other than reappear to reestablish the boundaries of life by means of death.

This is Agamben's implicit point about the total humanization of the animal coinciding with a total animalization of man. When humanity has been animalized and the animal humanized, which is to say practically that when a zone of indistinction is created between human and animal such that one can no longer distinguish man as human or man as animal, a situation may arise in which sovereignty reestablishes the borders of the human and the animal, which is to say, returns to establish what a proper human form of life looks like or how such a form acts. Agamben's fears over biometric *dispositifs*, ones deeply embedded, as he notes in the *Le monde* editorial, in nineteenth-century practices intended to block degeneration (and where there is degeneration, there is always eugenics, as we know), lie precisely in the possibility that such norms may in fact be used by a sovereign power (or here a governmental machinery such as that of the European Community, the specter of all of Agamben's recent reflections) to reestablish some sort of order in the ranks of proper and improper humanity.

The thanatopolitical drift of Heidegger's reading of technology, in which improper writing brings on a catastrophe in Being, becomes manifest. It does so when Agamben awards an implicit thanatopolitical value to Foucault's notion of *dispositif* by having it operate essentially as a metaphysical operator on Being, as technology did for Heidegger. The effects we know well. First, Heidegger's saving power is diminished or weakened, becoming instead a power over life. As I pointed out earlier, Agamben does so by drawing on *Gestell* (and with *Bestellen* and *Darstellen*) as a category that captures man, while simultaneously eliding the critique of communication that Heidegger reserves for improper writing in *Parmenides*. Second, by extending *dispositif* in this way to include technology and much more, Agamben essentially creates conditions in which proper and improper are no longer discernable when our gaze turns to the differences between human and animal. Although Agamben does not lay this out explicitly, as I'm attempting to do here, the problem with reading the *dispositif* in extended Heideggerian terms as technology is that it makes "care," which, for Heidegger, is decisive in making the human, difficult when not impossible to muster. Recall those pages I cited earlier from "Letter on Humanism":

> Where else does "care" tend but in the direction of bringing man back to his essence? What else does that in turn betoken but that man *(homo)* become

> human *(humanus)*? Thus *humanitas* really does remain the concern of such thinking. For this is humanism: meditating and caring, that man be human and not inhumane, "inhuman," that is, outside his essence. But in what does the humanity of man consist? It lies in his essence.[79]

The thanatopolitical for Nazism will be found in its ability to inflect care as a biological operator, to have substituted biology for ontology. That is the danger at which Agamben is hinting in his reading of Foucault; a Heideggerian care has, thanks to the emergence of *dispositifs*, been transformed into "total management of life," one that cannot do without the most important means for managing life, namely, death. The result is a truly staggering extension of death through the production of docile bodies.

Thanatopolitics and Ruined Oaths

Agamben takes up the relation of thanatos to an utterly improper working of *dispositif* from a different perspective in the recent *The Sacrament of Language*. In what Agamben subtitles an archaeology of the oath, he attempts to locate oath in relation to an anthropogenesis, that is, the origin of humanity as indelibly linked to oath as "an experience of language in which man was constituted as a speaking being," and so offers further insight into the depth of his thanatopolitical inflection of biopolitics.[80] The argument, which is as rich as one would now expect from Agamben, begins with the assumption of a performativity of the word that we know so well from Derrida's own reading of performativity. Agamben's attention, however, falls specifically on the term *sacrament* and its connections to the performativity of oath. Thus "the performative experience of the word is constituted and isolated in a 'sacrament of language' and this latter in a 'sacrament of power.' The 'force of law' that supports human society, the idea of linguistic enunciations that stably obligate living beings, that can be observed or transgressed derive from this attempt to nail down the originary performative force of the anthropogenetic experience, and are, in this sense, an epiphenomenon of the oath and the malediction that accompanied it."[81] From this originary separation between oath and malediction, collective life, Agamben will argue, today lives without the bond of the oath, thus dramatically altering the nature of collective life, or *bíos*, on one hand, and *zoē*, on the other. At this moment of the separation of *bíos* and *zoē*, the thanatopolitical appears in the disjunction, as he calls it, of what unified the living being to his own proper language through the oath. "On the one hand there is the living being, more and more reduced to a purely biological reality and to bare life. On the other hand, there is the speaking being, artificially divided from the former, through a multiplicity of technico-mediatic apparatuses, in an experience of the world that grows ever more vain, for which it is impossible to be responsible and in which anything like a political experience becomes more and more precarious."[82] Interestingly, Agamben will speak of a proliferation of spectacles, on one hand, in which empty words follow on empty words and, on the other, the legislative apparatuses that "seek obstinately to legislate on every aspect of that life on which they seem to no longer have any hold."[83] Thus, given the explosion of *dispositifs,* both spectacular and legislative, it is increasingly difficult, when not impossible, for the living being called man to speak, which means precisely to "take the word" *(prendere la parola)* and to "make it his own" *(farla propria).*[84] In this reappearance of the proper, we find Agamben rehearsing again Heidegger's ontological operators of technology, the proper and improper modes of writing that for Heidegger set man on the path to a catastrophe and to a possible moment of saving. And as was the case with *What Is an Apparatus?,* the limits of improper writing have been extended well beyond man's improper relation to the typewriter to include as well even the moment of speaking. Perhaps this is not so surprising because we often speak of the "act" of speaking, which recalls as well a proper relation to the word through the hand: it is an originary moment of speaking before the possession of I that Agamben is attempting to locate. Note, however, that for the distinction between a living being not in possession of his proper language and another speaking being who maintains an improper relation to the words spoken, what is required is that something similar to Heidegger's typewriter be evoked in order that the distinction between *zoē,* here the living being, and *bíos,* now the merely spoken (by spectacle), will be in effect. That technique is precisely the distinction between a benediction and a malediction as an ever-present possibility in the experience of the living being to be constituted as a speaking being. Indeed, Agamben will speak about that moment in precisely technical language—"Religion and law technicalize this anthropogenic experience of the word in the oath and in the curse as historical institutions, separating and opposing point by point truth and lie, true name from false name, efficacious formula and incorrect formula"[85]—or soon after, when he notes that in a "technical sense," something "said badly" becomes a malediction. We seem to have returned once again to the superimposition of language over *technē* in the Heideggerian sense, with the oath–malediction working as an anthropogenic operator that both constitutes the living being as speaking being and at the same time makes possible today a reversal toward a moment of prior separation, when the techno-mediatic conditions are such that no longer is she who takes an oath required to keep it, given the constant reversal of proper relations into improper caused by the workings of the *dispositif.*

In other words, according to Agamben, we find ourselves once again facing a massive sacralization of life such that at the moment when a speaking being attempts to make the word her own (and fails), then the only reality available to a living being thrown back literally onto life is the biological one characterized by its capacity to be killed and not sacrificed. At this juncture, I'm reminded of those unsettling pages of Rène Girard's in *Violence and the Sacred,* in which he argues that the act of demystification of the mechanism of the sacred actually leads to greater violence:

> The act of demystification retains a sacrificial quality and remains essentially religious in character for at least as long as it fails to come to a conclusion. . . . In fact, demystification leads to constantly increasing violence, a violence less "hypocritical" than the violence it seeks to expose, but more energetic, more virulent, and the harbinger of something far worse—a violence that knows no bounds.[86]

Something similar may be at work in Agamben's reading of the diminished role of oath for collective life—he appears to imply that with the end of the possibility of oath as a mode of properly speaking (though it is clear that such a possibility is denied from the origin, according to Agamben), it becomes possible to witness the "sacrificial effects," as Girard describes them, of just such a process. For Agamben, those effects are principally in the extension of *homo sacer* to ever greater numbers of those who, in saying *I,* no longer enjoy a proper relation to language.

This moment of the contemporary impossibility of swearing an oath, or better, properly being the subject of an oath, is at the heart of the deeply thanatopolitical reading Agamben gives across almost four hundred pages of *The Kingdom and the Glory.* It is thanatopolitical because, as Agamben himself notes, in the instant in which individuals and collective are destined to give an oath in vain, a space opens up in which "politics can only assume the form of an *oikonomia*, that of a governance of empty speech over bare life."[87] No clearer statement will be found to demonstrate that for Agamben, governing today is a form of thanatopolitical governance because there every oath spoken in the name of life will always be an empty one. The abandoning of some lives to their biological reality not only occurs because of a depoliticization we know so well from Hannah Arendt but is put into practice given the fundamental caesura between a living being and a speaking being, whose oath toward all forms of life is one to which a government of *oikonomia* no longer maintains any proper bonds.

Moreover, this homology between *oikonomia* and bare life also allows us to see just how fundamental the secularization of theological forms in contemporary governance is for Agamben. On the one side of what Agamben will call the *dispositif* of *oikonomia* in *The Kingdom and the Glory* is the Kingdom and, on the other, the Government, and between them perpetual conflict, with one side dominating over the course of modern history. He will then observe how "liberalism represents a tendency that pushes to the extreme the supremacy of that 'immanent-government-stomach' pole to the point of almost eliminating the 'God-transcendent-kingdom-brain' pole. . . . And when modernity abolishes the divine pole, the economic that will derive from it is not in any way freed from its providential paradigm."[88] *Dispositif, machine, paradigm*: these are the figures that populate Agamben's thought, and here, just as across so much of his work, we find them working deeply to introduce death into life. What he has done in his reading of the economic in *The Kingdom and the Glory* is to provide his version of Foucault's *The Birth of Biopolitics*, elaborated as the remnant of a theological economy that has eliminated God but not providence. The result, as in *The Sacrament of Language*, as before in *What Is an Apparatus?*, is the removal of the stops in the sacralization of life through the proliferation of *dispositifs*. And in this regard, the appearance of the *dispositif* of *oikonomia* in modernity would signal an immense intensification of the lethal effects of the *dispositif* in general because for Agamben, it is through the *dispositif* principally that processes of subjectivization take place. In other words, the providential paradigm that still holds in modernity is precisely the one that continues to make *homo sacer* of those governed, that continues to make available for those with eyes to see the falling away of every personal identity, which cannot be thought independently of the thanatopolitical.

Returning again to *What Is an Apparatus?*, Agamben speaks of these processes of subjectivization in terms of a "dissemination that pushes to the extreme the masquerade that has always accompanied every personal identity."[89] When taken together with Agamben's reading of the impossibility of giving an oath and being held to it, we can make out the terrifying effects of a liberal governance, the dismantling of individual responsibility as well as the abandonment of forms of personal life. In their place is the spectacle of empty words spoken by figures whose masks are clear to all except themselves. If these masked subjects do not enjoy a proper relation to their own words nor to their own personal identities, what remains is merely the biological reality of their bodies, their DNA, their fingerprints—what biologically distinguishes them from others.[90]

Again, such a view of the biological distinction between living beings as the sole feature that separates human beings rehearses, too, those moments of Heidegger's perspective on mankind, communication, and proper writing. With the turn toward the economic as the principal mode for making sacred vast numbers of humanity, however, Agamben intensifies the role of improper writing as now the chief characteristic of this moment of unmatched economic governance. In no period more than today does the economic govern lives in providential fashion. Agamben not only superimposes the proper and improper forms of writing over proper and improper speech through the introduction of the *dispositif* of the oath but in fact has made the *dispositif* the mechanism by which a modern *oikonomia* comes into being. And no *dispositif* is more powerful than the economic to "capture, orient, determine, intercept, model, control, or secure the gestures, behaviors, opinions, or discourses of living beings."[91] The providential machinery that Agamben sees producing bare life today in the West looks remarkably similar to the machinery of proper and improper writing that "enframed" mankind, that made her and him ready at hand, and that made them resemble everyone else, as the typewriter did for Heidegger. If there is a difference, it resides in Agamben's having provided an astounding genealogy for the economic in the theological equivalent of kingdom and government.

Affirming the *Dispositif*

Another important contribution to understanding the seemingly inevitable inscription of contemporary philosophy in the thanatopolitical will be found in Italian philosopher Robert Esposito's most recent work,

Terza persona: Politica della vita e filosofia dell'impersonale, as well as an important essay that responds to Agamben's *What Is an Apparatus?* titled "The Dispositif of the Person." I have had occasion to discuss Esposito's perspective on biopolitics in a variety of different venues. In most of these, I've focused on his *Bíos: Biopolitics and Philosophy* as well as his earlier *Categorie dell'impolitico.* There Esposito lays the foundation for a politics of life as opposed to a thanatopolitics of death. In the pages remaining, I want to recapitulate Esposito's perspective on the thanatopolitical, especially as it relates to his reading of Heidegger as well as his understanding of the biopolitical effects of the term *dispositif* through these more recent works. Some of the value of Esposito's response will be found in having found the means to reinsert an affirmative aspect to contemporary perspectives on the *dispositif.* He will do so in ways that Agamben does not by drawing on those sections of Deleuze's essay on the *dispositif* and employing them so that it becomes possible to imagine a radically negated thanatopolitics. Here I have in mind not only those sections in which Deleuze, formulating a different genealogy of the *dispositif* removed from a Heideggerian ontology of being, associates Foucault's thinking to a tradition of "intrinsic aesthetic modes of existence" that will run from Spinoza to Nietzsche to Georg Simmel.[92] Of particular interest are those pages in which Deleuze notes how "modes of existence have to be assessed according to immanent criteria, according to their content of 'possibilities,' liberty or creativity, without an appeal to transcendental values" and then links such moments to a later, more ethical and less political Foucault in his allusion to "aesthetic criteria, which are understood as criteria for life" and which "replace on each occasion the claims of transcendental judgment with an immanent evaluation." Could the outcome be, Deleuze asks, the elaboration of "the intrinsic aesthetic of modes of existence as the ultimate dimension of social apparatuses [*dispositifs*]?"[93] We know how Agamben will answer the question: thanatopolitically, with those modes of existence translated as bare life, the desubjectified subject and the aesthetics of testimony that emerge as integral to his perspective on everything from life to literature to politics. For Esposito, what will count most in Deleuze's reading of the *dispositif* is how the modes of existence that Deleuze links to possibility and freedom may be brought together most forcefully in a conception of the impersonal.

At first glance, this does not appear to be the case. Esposito does not lay out anything like a methodological manifesto of the sort Deleuze and

Agamben offer. Rather his perspective only becomes visible when elaborating the category of the person as a *dispositif* intimately associated with contemporary forms of thanatopolitics. To see how, a simple compare and contrast between Agamben and Esposito will suffice. Essentially, where Agamben flattens Deleuze's modes of existence into desubjectified subjects and those who chronicle that traumatic process, Esposito prefers to distinguish between subject and person. The reasons for this change of emphasis to person—and we should note that Agamben rarely makes a distinction between depersonalization and desubjectification—are many. First, there is, of course, Esposito's attempt to think an affirmative biopolitics through a philosophy of the impersonal, which, of course, the notion of subjectivity (and desubjectivity) fails to do: no such category of an "im-subject" is readily at hand, and furthermore, in classical models of the subject, one is or one is not a subject. Second, a quick glance at the Internet will confirm that the notion of person occupies a much more important role than the subject in current debates ranging from euthanasia to abortion to the mapping of the genome.[94] My impression is that the term *subject* remains wedded to a one-dimensional view of biopolitics as always already inscribed in a "sovereign biopolitics," and hence its use value for a different inflection of biopolitics offers little. Person and its impolitical form that go under the name of the impersonal offer a much larger purview for philosophy to come to grips with what is most pressing today—thinking life beyond merely *zoē* and *bíos*.[95]

A final word of introduction: for Esposito, the *dispositif* of person has long been in operation—his reading of Roman law, which I will discuss shortly, confirms that since its origins, the notion of person has had effects of ordering different registers of persons, those "modes of existence" that Deleuze sees as the groundwork for a possible philosophy of the *dispositif* of the future. Here Esposito's task, however, is to show how "only through the lexicon of the person does a notion like that of human rights become conceivable and practical"—and here Roman legal conceptions of person are key.[96] At issue is that only those beings awarded the status of person enjoy the fundamental rights of being human; in other words, another category sits astride the notion of human rights, the subject of human rights, as well as juridical subject, and that is person. Essentially, Esposito will read the subject, be it the Foucauldian subject or the juridical subject, as a "personal subject" that today is to be thought as separate from the human being. In terms of the thanatopolitical drift I noted in Heidegger's discussion of technology and ontology, Esposito's adoption of *dispositif* in speaking of person both gestures to the separation between proper and improper that Heidegger sees operating in forms of modern writing and seeks to move past it by employing another category. Two moves, therefore, characterize Esposito's thinking of the thanatopolitical and technology: on one hand, the implicit use of Heideggerian notions of proper and improper in the workings of *dispositif*, on the other, the need to overturn them so as to construct a politics *of* life, as he subtitles, *Terza persona*.

What distinguishes the *dispositif* in Esposito's reading in both "The Dispositif of the Person" and *Terza persona* is his willingness to see in the working of the *dispositif* not just the processes of subjectification but instead the vehicle by which a regime of personhood is instituted. Again, it is the *dispositif* that institutes the person and the process of depersonalization. This inflection away from subject depends in large measure on a subtly anachronistic reading of both Roman jurisprudence and the indirect deconstruction of Agamben's *oikonomia* through a reading of the notion of Trinity and person in Christian theology. As a way of drawing attention to Esposito's inflection of the thanatopolitical drift I noted in Heidegger, and of course, in Agamben, toward a power *of* life, I want to cite two passages from Esposito. The first comes from "The Dispositif of the Person," in which Esposito distances himself from what we might consider a classic reading of *dispositif* in Foucault. He does so by noting how the Christian idea of person encapsulated in St. Augustine's dictum that "each individual man, who is called the image of God . . . is one person [*secundum solam mentem imago Dei dicitur, una persona est*]" can help us position *dispositif* in terms of its performative value today: "Already here, and in a formulation unparalleled for its dogmatic clarity, the Christian idea of person adheres to a unity that consists not only of a duplicity, but such as to subordinate one of the elements to another to the point at which it expels the other from the relation with God." Continuing, Esposito writes,

> But distance from God also means the weakening or the degradation of that humanity that only takes on its proper and ultimate truth thanks to a relation to the Creator. For this reason the need in mankind to provide for his bodily needs can be defined by Augustine as a "disease."[97]

The point of these readings arrives immediately after in terms explicitly informed by Simone Weil's notion of the impersonal: "what does not belong to man is not properly human in the specific sense that it is the impersonal part of his person."[98] The passage is remarkable for the eloquence with which Esposito grafts a Heideggerian division of proper and improper into his interpretation of Augustine as well as a distancing from the Heideggerian motor of proper and improper through the impersonal.[99] To see how Esposito deploys Augustine's argument against Agamben, let's recall the passage that immediately precedes the one Esposito includes here. Augustine writes:

> For man, as the ancients defined him, is a rational mortal animal. These things, therefore, are the chief things in man, but are not man themselves... But if, again, we were so to define man as to say, Man is a rational substance consisting of man and body, then without doubt, man has a soul that is not body, and a body that is not soul. And hence these three things are not man, but belong to man, or are in man.[100]

What Esposito latches on to is the implicit difference between what belongs to man, those elements that will characterize him, namely, rational substance, soul, and body, and what is man. The *dispositif* of person allows Augustine to bring together the "chief things" and the person of man, which is to say, what separates man from what properly belongs to him. To the degree the three things mirror the Trinity (Father, Son, and Holy Spirit), Trinity names the totality of God and, by extension, the category of person that subsumes man. The explicit critique of Agamben is clear enough—for Agamben, the notion of the Trinity served to allow Tertullian and Ippolito to say that "God . . . is certainly one; but as for his *oikonomia,* that is with regard to the mode with which he administers his house, his life and the world that he created, he is instead threefold."[101] Left unanalyzed in Agamben's account of *oikonomia,* however, is the role of the *dispositif* of person, which, rather than separating being and action, or "being and praxis" in God, as Agamben says, instead separates in man those things that together make up his person from man proper.[102] Person is the *dispositif* that separates man from the three things but also the mode by which God and man are thought together (as well as distinguished from one another). Here, then, would appear to be a moment originary with the birth of *oikonomia*: the emergence of the *dispositif* of the person as well as its resemblances with the notion of Trinity.

The full import of Esposito's articulation of person and subjectivity now comes clearly into view. By noting how subject and object were, until Leibniz, indistinguishable from one another, Esposito will essentially make homologous the process whereby one is subjected to the process of being made an object. The performance of the *dispositif* of the person is situated there:

> Person is precisely that which, dividing a living being into two natures of different qualities—the one subjected to the mastery of the other—creates subjectivity through a procedure of subjugation [*assoggettamento*] and objectification. Person is that which renders a part of a body subjected to another to the degree in which it makes of the second the subject of the first.[103]

To be a person is not to live the separation between proper and improper but rather to be divided so as to make possible the subjugation of one part to another. The process of subjugation and thus of constructing the subject cannot be thought apart from the ordering of two different natures. This is the crucial node for Esposito for thinking the thanatopolitical, which he will do in extended form in *Terza persona,* especially through the category of person in Roman jurisprudence. Yet consider something else that is implicit in the preceding passage. We see not only the zone of indistinction that arises between subject and object, with the result that to be subjected in the *dispositif* of person also means to be objectified, but also the superimposition of objectification with the body. Esposito's reading of person is both compelling and troubling precisely because he is arguing that to be a person means quite literally to be divided in half so as to make the other the object of the first. Through a semantic chain that will join object to slave to animal, Esposito will then posit a fundamental fissure in the ontology of person that inscribes the animal within it and simultaneously outside of it:

> Man is a person if and only if he masters the more properly animal part of his nature. He is also animal but only so as to be able to subject himself to that part which has received as a gift the charisma of person. Of course not everyone has this tendency or disposition to deanimalize. The degree of humanity present in all will derive from the greater or lesser intensity of deanimalization and so too the underlying difference between he who enjoys the full title of person and he who can enjoy it only if certain conditions have previously been met.[104]

The *dispositif* of the person entails an ever shifting line, a *dis-porre* between deanimalization and subjectivity. Much depends on the gift of charisma from God, which moves man ever closer in the continuum between animal and human toward the human. The less one's nature is disposed toward animalization, the more one enjoys the title of person and hence the more human one is. What Esposito has unearthed, seemingly, is a subtle layer between the human and the animal that before either went unremarked or was subsumed into the larger category of either subjectification or *dispositif.* A number of questions will be raised, some of which will concern Heidegger and the thanatopolitical drift we noted in "Letter on Humanism," in particular (as that is the text that Esposito seems to be implicitly invoking throughout his essay). I will turn to some of these questions shortly, but before I do, I did want to look at one final detail of Esposito's analysis, which concerns the role that charisma plays in the *dispositif* of the person. From the Greek, *kharisma* names the "gift of grace" and is, of course, an integral part of an early Christian lexicon. Equally, the secular form of grace plays an important role in Max Weber's analysis of different forms of modern leadership. In this regard, Weber defines charisma as "a certain quality of individual personality by virtue of which s/he is set apart from ordinary people and treated as endowed with supernatural, superhuman, or at least exceptional powers or qualities. These as such are not accessible to the ordinary person, but are regarded as divine in origin."[105] Weber's definition helps us locate the fundamental role grace plays in modern conceptions of what is meant by person. But where Weber sees charisma as a fundamental operator in the *dispositif* of person, one that separates ordinary people from the extraordinary individual—a formula that will mark so much of modern forms of personalist philosophy and that appears to invoke the terms of proper and improper, namely, a proper individual and improper and ungifted people—Esposito transfers that difference that separates the subject and body itself.[106] Consider that the divine gift of grace does not fall to both the animal and the human but rather is given to that part that subjects, which is to say, has power; indeed, it would appear that for the dominating part to subject the former, it must be awarded with grace. Thus the interval between the personal subject and the human being of the twentieth century that the *dispositif* of person subsumes will also depend on one part enjoying a closer, more intense relation with the divine or its modern substitutes (the nation for totalitarianism, individual "charm" in liberal democracies today). And if we consider as well that many nonhuman actors, corporations in particular, are considered persons under the law, a whole series of semantic chains emerge that would link an earlier form of grace to the ultimate awarder of grace today, namely, the market.

With more time, it would certainly be profitable to read Foucault's critique of neoliberalism in this key.

In much of what I have drawn from Esposito's essay, my emphasis has fallen on the early Christian roots of the *dispositif* of the person. And while that has, I hope, made clearer our own relation to the *dispositif* of person today, I have given short shrift to the fundamental role that Roman jurisprudence plays as well. If we are to register Esposito's analysis of the thanatopolitical, however, we should at a minimum set out the relation between the thanatopolitical and the *dispositif* of the person in Roman law. Without being able to enter all the details of Esposito's rereading of Roman law and its anachronistic presence in our contemporaneity, what emerges from his reading is, in fact, the mobile status of the person—its capacity to create thresholds among not just subject and desubject but a host of persons who move from one register of person to another, creating vast zones of indistinction among them. Here the reference to Agamben is explicit enough, but what sets Esposito's discussion apart is the superimposition of slavery and thing and suturing the concept of person to join opposites (or allowing them to be broken apart). In terms of the slave in Roman law, we recall that what characterized the slave principally was the middle status occupied between person and thing such that the slave was defined as a living thing and as a reified person. The status of the slave as thing will allow Esposito to radicalize the very nature of person as literally involving the creation of new slaves as a way of guaranteeing one's own personhood in Roman antiquity:

> Here then is the perpetual oscillation between the extremes of the person and the thing that makes one its opposite and the background of the other—not only in the general sense that the definition of man-person emerges negatively from that of man-thing, but in that more pregnant sense that to be fully a person means to maintain or push other living individuals to the border with the thing.[107]

Esposito introduces another two figures from antiquity, *manumissio* and *mancipatio,* to name the continual movement of personalization and depersonalization, but what counts is the place awarded "thing," which, translated into Heideggerian terms, appears as the improper abode for being, for the depersonalized person.[108] Thus to be a person in Roman antiquity, others had to be pushed across the threshold of the person so as to become less a person. It also meant that those who were already nonpersons, slaves, had to stay that way. Although Esposito goes out of his way to deny any compatibility between this Roman *dispositif* and a semantics of the person today—the figure of Hobbes is crucial here for the Roman *dispositif* of person's modern translation into person and mask—we fail to register the radicality of Esposito's superimposition of Roman law and our contemporaneity if we don't allow for the possibility that today, to be a person involves maintaining other living beings as things or, indeed, pushing them over into the realm of the thing. In my view, this is a fundamental part of how neoliberal governmentality works today; indeed, it would be productive to recast these reflections of Esposito's in his earlier terms of immunization and community, for what they entail is not simply the sacralization of vast swaths of populations but effectively the constant rezoning of populations as persons along a sliding scale toward the thing.[109]

Directing Esposito's reflections on personhood toward a critique of neoliberalism is one way of examining the effects of the *dispositif* of personhood today. Equally, the *dispositif* of the person figures as an essential motor for twentieth-century thanatopolitics. And yet Esposito will not make the move we might expect, namely, that the Nazis, to maintain a notion of person, will turn millions into nonpersons, who can then be killed. The reasons for not doing so are clear. Not only would this be a misreading of Nazism's fundamental biocratic features but would collapse Roman law into a form of thanatopolitics avant la lettre. The point of introducing the Roman conception of juridical person into a discussion of biopolitics is, beyond marking points of contact that are productive for distinguishing our contemporaneity from antiquity, to help us distinguish in ways we have not the distance between liberalism, for instance, and Nazism. My own impression, different from Esposito's, is that this distance grows ever smaller under a neoliberal governmentality. Be that as it may, how are we to think thanatopolitics through the person? Thanatopolitics will not consist primarily of the attempt to turn persons into things (or people into populations, pace Foucault and Agamben) but rather to crush the person and thing, to make them coextensive in a living being. This thing in thanatopolitics is taken to be the body, the biological material that assumes both the person and the thing, the person as a biological thing. This biological thing that lives belongs to the person as her property. This was the liberal vision of Locke and Mill in which "the predominance over the object is not established by the distance that separates it from the subject, but by the movement of its incorporation."[110] In Nazism, these bodies, as Esposito shows, belonged to the Führer, who could do with them what he wanted.[111] It appears, then, that a wide gulf separates the *dispositif* of the person in liberalism and Nazism such that the thanatopolitical drifting that we noted first in Heidegger and then as complete in Agamben is here lessened.

Yet Esposito's reasoning is more complex than that. Not wanting to collapse Nazism and liberalism together into some unworkable conception of the thanatopolitical does not stop him from noting a certain symmetry, if not a radical superimposition and radical collapsing of cultural, ethical, and political dimensions, between the two. What they both share is "a productivist conception of life," which, in the case of Nazism, was "made functional to the superior destinies of the elected race and in the other to the maximum expansion of individual liberty."[112] Liberalism and Nazism avail themselves of the suturing possibilities offered by the *dispositif* of the person for different ends, but what cannot be in doubt is that they both utilize the same *dispositif.* The separation that the *dispositif* of the person institutes, then, between person and what belongs properly to the body is what allows one, for instance, to donate organs, to oversee and manage one's body, one's own corporeal capital, and to do with it what one will in the name of an expansion or better production of individual liberty. Such a production is premised on the possibility of administering forms of death on oneself, on the biological thing that I administer—or, in the case of Nazism, to manage the biological health of the body politic by administering death.[113] Although Esposito does not directly evoke Foucault here, he clearly does have the Foucault of *The Birth of Biopolitics* in mind, especially those pages in which Foucault posits, almost approvingly, the capacity of neoliberalism to allow individuals to profit from their own individual biopower.[114] So, too, do these pages from Esposito evoke others from his earlier work on the impolitical. Many come to mind, but especially important are those in which he distinguishes between *potere* and *potenza* (which reappears in *Bíos* in a different form):

> The same semantics of the body that has a liberatory, antinomic function evidently implies a process of potentializing [*potenziamento*]. The body opposes the lexicon of power [*potenza*]—multiplicity, change, exteriority—to that of power [*potere*]. But this is the reason that the subject is

> augmented, filling up the body, and so, inevitably, also the power that constitutes it as such. In this sense, therefore, power [*potenza*] cannot be withdrawn from its destiny of becoming powerful [*potere*].[115]

With this in mind, we might say that in liberalism and, more intensely, in neoliberalism, the *dispositif* of the person represents a powerful mode by which an individual harvests his own biopower through a process of potentializing his second nature—to use the body as a biological material or thing through a reverse zero-sum game, thanks to a split between natures brought on by the *dispositif* itself. That is exactly the sense with which Esposito speaks of a "liberatory function"—here, though, in the service of neoliberal expansion of individual biopower.

The implicit impolitical features of neoliberal biopower and governmentality merit a much longer discussion, but as the preceding passage makes clear, neoliberalism asks, as Foucault knew so well, not just how to limit the exercise of the power of governing but also how to create conditions in which one might draw forth and profit from one's own body. The decisive condition for that will be making sure the body and the person are never coextensive but also that a person's capacity to increase her biopower will become the primary means for determining how fully she is a person. The reappearance of the Roman separation between *persona* and *homo* will come to mind at this juncture, as will those writers, such as Hugo Engelhardt and Peter Singer, who argue along the following lines:

> This use of "person" is itself, unfortunately liable to mislead, since "person" is often used as if it meant the same as "human being." Yet the terms are not equivalent; there could be a person who is not a member of our species. There could also be a member of our species who are not persons. . . . In any case, I propose to use "person," in the sense of a rational and self-conscious being, to capture those elements of the popular sense of "human being" that are not covered by "member of the species Homo sapiens."[116]

As we know, the decisions about how to employ or dominate this other nature—call it biology, the animal, or what you will—are increasingly thought through the market. João Biehl has movingly described these processes in terms of social abandonment in neoliberal Brazil. He captures well the function of the *dispositif* of person today: "You shall be a person there, where the market needs you."[117] Furthermore, attempts to "fill up" the person with this secondary material not only consist in making a person stronger, faster, and more intelligent but also its reverse, through the biomedical power of pharmaceuticals, so that a person can essentially put his body—his animality—on hold. Biehl writes with regard to AIDS in Brazil, "They know they are trapped between two destinies: dying of AIDS like the poor and marginal, that is *animalized,* and the possibility of living pharmacologically into a future, thereby letting the animal sleep and preventing it from consuming the flesh."[118] As Biehl describes it, personhood today is made or unmade in an instant, creating regimes of persons, semipersons, nonpersons, and antipersons, whose function is to make it easier for them to be abandoned. The role of medicine is absolutely crucial here, for it lays out in front of each person, in ways that give the lie to the truth of the market, the possibilities for future personhood.

I have introduced Biehl's work here not only because it is one of the most powerful dispatches from the front of neoliberal governance but also because he allows us to register in real terms the stakes of Esposito's reading of liberalism. The question is not whether Esposito's perspective on the *dispositif* of the person offers a way of contextualizing contemporary thanatopolitics—obviously it does—but rather how we wish to think of these recent reflections from Esposito on liberalism and an implicit thanatopolitics and the Heideggerian drift toward the thanatopolitical that is the subject of this discussion. Undoubtedly, Esposito, much more so than either Agamben or Sloterdijk, is deeply conscious of the problems that any appropriation of Heidegger creates, especially as regards the transposition of *eigentlich* and *uneigentlich* (proper and improper) into a perspective of life. Here Esposito's fundamental deconstruction of these terms in a larger context of community merits attention:

> The purpose of community, if it is admitted that one must speak of purpose, cannot be that of erasing community's own negative, that is, of bridging over the interval of difference, of achieving community's own essence, and not because community fails to aspire to be properly its own. The reason instead concerns that what is properly ours *(il nostro propio)* doesn't reside anywhere else except in the knowledge of our "impropriety."[119]

Any attempt to grab hold of the improper—here the context is the community, but it also includes all attempts that concern weeding out the improper, making it the object of a proper "care," to borrow the Heideggerian term—will fail because the proper does not reside apart from the improper. In different words, the distinction between proper and improper collapses under the weight of the proper's own improper content. Heidegger, as Esposito goes to great pains to show, argued thusly in his early conception of *Dasein*: "Improper in this case is not other from proper, but is Dasein itself understood in its own proper impropriety."[120] And yet Heidegger's precise failure was to have forgotten the radical deconstruction of the improper origin of proper; this, Esposito says, will be the terms of Heidegger's Nazism, though we may substitute thanatopolitics here: "This and nothing else was Heidegger's Nazism: the attempt to address directly the proper, to separate it from what is improper, and to make the improper speak affirmatively the primogenital voice; to confer upon it a subject, a soil, and a history."[121] Esposito does not deny the usefulness or the appropriateness—to stay with the figure of the proper—of the terms *proper* and *improper*, for thinking either community or an affirmative biopolitics, but only so long as they become the means by which the entire *dispositif* of proper–improper is seen as improper, as escaping any privileging of a proper body, a proper nation, a proper person.

A series of other questions arise at this point: how is the separation between various degrees of persons instituted by the *dispositif* of person to be remedied? Is the separation between proper and improper always on the road to the thanatopolitical? In what sense does neoliberalism attempt to reinstitute a separation between proper and improper under cover of the expansion of an individual's freedom? What are the risks we run by superimposing a Heideggerian division between proper and improper over Roman law's division between *persona* and *homo*? Furthermore, once we have deconstructed proper and improper, what do we make of the connection between proper and improper care for the other and the regimes of personhood that dominate the landscape of life today? Does not a discussion of the mobile thresholds instituted by the *dispositif* of person call forth different regimes of care? I am reminded again of Biehl's anthropological description of Brazil as being mapped according to zones of abandonment. There he distinguishes between persons who are the objects of affection and those others who are not, who are only remembered. When we fail to separate proper from improper, what happens to our notions of justice that are premised on an abiding distinction between proper and improper?

All these questions demand attention. That we can ask them depends precisely on the workings of the *dispositif* of the person. In my view, Esposito believes that the Heideggerian separation between proper and improper, which heralds the institution of the modern *dispositif* of the person, inevitably calls forth the drift toward the thanatopolitical. It is as if a power stronger than thinking the origin of the proper in the improper's operation here—that the power of the *dispositif* of the person, whose intensity we now understand, given its symmetry with Roman jurisprudence—won't go willingly into the night. The earlier moments I noted in Heidegger's thought of proper and improper writing as calling forth danger, and Western man, who must be saved, those moments linked to cellular phones, gadgets, all forms of communication that flatten distance and difference between individuals—those moments, in Esposito's perspective—draw less attention for their potential thanatopolitical drift than they do either in Agamben or Sloterdijk. This diminishment of technology for understanding the current biopolitical moment is one that requires explanation. For Esposito, modern technology may actually play a role in pushing the thanatopolitical inflection of contemporary biopolitics toward its tipping point; that is to say, when the separation between proper and improper has reached its maximum point of extension, then the power of knowing the improper content of the proper will emerge as increasingly self-evident. Thus there may well be a powerful project of Enlightenment in Esposito's work, a vector that emerges near the end of *Bíos* and now in *Terza persona.* It is what he calls the politics of life, thought not through the personal, for all the reasons we know: the implicit division of man's nature in the person allows the biopower of the individual to be more easily captured by either the State or, today, the market through the mediation of the neoliberal subject intent on augmenting his own biopower. If there is an emphasis to be found on technology and thanatos in Esposito, it concerns the effects of technology on immunization, especially those modes of surveillance and registration of data whose effect will be to augment calls for protection from the very means of technical immunization itself. This again suggests a sort of tipping point, though that point is never thought alone as a danger—or rather the thanatopolitics of biometric scans, for instance, does not inevitably lead to the sacrifice of *homo sacer,* be it of their political life, their *bíos,* or their unstoppable march toward desubjectification. This point of no return in Esposito's terms may actually lead to a reversal of the current terms in which thanatopolitics, the *dispositif* of the person, and technology are currently configured. Before I tell you the name that reconfiguration goes by, let's recall that this capsizing of a political lexicon does not involve a resocialization of biopolitics of the sort that Antonio Negri with

Michael Hardt urged forward in *Empire* and *Multitude*, which is akin to pouring the new wines of rhizomatic networks and singularities into old flasks that had contained the earlier models of communication, sociality, and interaction.

The reversal Esposito has in mind concerns, rather, elaborating an oblique angle on the *dispositif* of the person, a perspective that draws on the grammatical category of the third person to create a breach in the regime of the person.[122] The thought of Simone Weil is deeply important here, especially those pages that Esposito cites with regard to the impersonal. After citing Weil at length, especially the famous passage "so far from its being his person, what is sacred in a human being is the impersonal in him. Everything which is impersonal in man is sacred, and nothing else," he will then define the impersonal this way: "The impersonal is not the simple opposite of person . . . but something that interrupts, either that is of the person or in the person the immunitarian mechanism that places the I in the inclusive and excluding circle of us."[123] This sentence, which invokes Esposito's earlier work on immunity, makes it clear just how he sees the *dispositif* of the person operating today. The separation enacted by the *dispositif* of the person does not just separate (persons from each other and my own person from my biological material) but also contains within it a function whereby the person both belongs and does not belong to a plurality. To think the impersonal, but more important, to act impersonally, breaks with the proper and improper—breaks with what is considered ours and what is considered mine. We recognize in these formulations a shorthand for the thanatopolitics of totalitarianism that attempts to make what is ours ever more ours and liberalism, in which "I" am given enormous freedom in determining how best to use or consume what is properly mine (and what is more properly my own in neoliberalism than the body I inhabit as a person?). Clearly I am pushing Esposito's thought here in directions he does not explicitly acknowledge; his object is less a distancing from thanatopolitical than it is a laying of the foundations for an affirmative biopolitics. Citing Weil in this instance is intended to create conditions in which a radical opening of the human community to the impersonal can occur, an opening sundered—perhaps from its inception—by the *dispositif* of the person; so, too, to reestablish the primacy of obligations over rights that implicitly recalls and extends a form of republicanism.[124] Yet it is not difficult to see in this juxtaposition of "right" on one side, as belonging properly to the person and personal, and "justice" on the other, as concerning the impersonal, a wide-ranging critique not only of thanatopolitics but of the entire biopolitical *dispositif* that is based on the conception of proper.

Let me try to tie up the loose ends of Esposito's engagement with and distancing from Heidegger's discussion of technology as primarily proper and improper writing. First, consider how far removed Esposito's and Weil's reflections on the impersonal are from Heidegger's own, as set out in *Being and Time*. Discussing the relation between *Dasein* and *die Öffentlichkeit* (public realm), Heidegger writes, "We enjoy ourselves and take our pleasures as *they* do; we read, see, and judge works of literature and art as *they* do; but we also shrink back in revulsion from the 'masses' of men just as *they* do; and are '*scandalized*' by what *they* find shocking. The they, which is nothing definite, and which all are, though not as the sum, prescribes the Being of everydayness. The 'they' has its own ways in which to be."[125] As David Farrell Krell notes,

> Heidegger argues that the public realm—the neutral, impersonal "they"—tends to level off genuine possibilities and force individuals to keep their distance from one another and themselves. It holds *Dasein* in subservience and hinders knowledge of the self and the world. It allows the life and death issues of existence to dissolve in chatter which is the possibility of understanding everything without prior dedication to and appropriation of the matter at stake.[126]

The result is *Dasein* held hostage once again to the social leveling associated with improper writing. This superimposition of improper writing and the impersonal brings us full circle in our discussion of the drift toward the thanatopolitical in contemporary thought; it indicates that perhaps the thanatopolitical may have found a back door in Esposito's thinking not because of its continued inclusion in a *dispositif* of the person but because of the threat it poses for individual identity. In other words, the third person singular or plural may remain, in fact, inscribed in an improper relation of Being to man. The typewriter for Heidegger makes everyone the same, which could be seen as producing an impersonal writing. If that is the case, then any hoped-for escape from the including–excluding *dispositif* of person comes up against the threat that improper writing poses to being. Said differently, in adopting the impersonal, we put ourselves in the position of a writing that, for Heidegger, was already on the way to the thanatopolitical (since it withdraws man from a proper relation with *Dasein* while also separating him from others). The impersonal, seen from a Heideggerian perspective, in no way protects man from a destiny linked to the increasing domination of improper writing.

My own impression is that Heidegger and, with him, Agamben willfully exaggerate the risks of technology and improper writing. Must we inevitably read Heidegger's danger or Agamben's devastation in terms of the complete annihilation of life? Perhaps we can reverse the question of the impersonal and the improper and see that we need not reject outright an impersonal perspective because of its associations with an improper writing that levels distinctions among persons. In the current moment of personhood and the ever increasing and accelerating moves from person to semiperson and back, a perspective, or better, a practice that would also include a passage into the impersonal could create a space in which Esposito's project of "common law" becomes meaningful (or *pace* Agamben, a "common use"). Indeed, there is precedence for just this kind of impersonal practice in the later Foucault, especially the courses Foucault offered in which he elaborates an art of governing oneself by explicitly invoking the Heideggerian notion of care. In an interview published before his death, Foucault inscribes this care of the self in the horizon of liberty. In a quotation that merits a much longer citation than I can give here, Foucault distinguishes between liberation and liberty. It is one thing to be liberated from a colonizer and another to practice liberty:

> According to this hypothesis, all that is required is to break these repressive deadlocks and man will be reconciled with himself, rediscover his nature or regain contact with his origin. . . . I think this idea should not be accepted without scrutiny. But we know very well . . . that this practice of liberation is not in itself sufficient to define the practices of freedom that will still be needed if this people, this society, and these individuals are to be able to define admissible and acceptable forms of existence or political society.[127]

One can hear in Foucault's words a privileging not of communication (the emphasis on the individual and forms of existence is proof enough) but of something like expressionism, in the sense that Deleuze speaks of it, impersonally: "And what takes the place of communication is a kind of expressionism. Expressionism in philosophy finds its high point in Spinoza and Leibniz. I think I've found a concept of the Other, by defining it as neither an object nor a subject (an other subject) but the expression of a possible world."[128] The impersonal would emerge concurrently as a practice of critical immanence and of liberty as the encounter impersonally with other possible worlds that in no way diminish Being or lead to its withdrawal.

Death, of Its Own Accord

I have attempted in this chapter to mark the principal features of the thanatopolitical in Giorgio Agamben's and Roberto Esposito's thought. Agamben tends, in a number of recent writings, to augment the power of improper writing, given the ease with which he crosses it with a Foucauldian notion of *dispositif.* The result is that the soon-to-be-arriving catastrophe will be profoundly catastrophic thanks to the capacity of the *dispositif,* what we might call, at this stage, an utterly improper *dispositif,* to aid in the withdrawal of Being from beings. A related result is that human beings on a global scale have lost some of the principal features that make them human, becoming, to a large degree, all too inhuman. One of the principal figures of the thanatopolitical in Agamben's thought is that of desubjectivization brought on precisely by the proliferation of *dispositifs*; in the process, bare life has been extended across the globe. Agamben, in fact, in a recent text, links this extension of bare life to all to what he calls "democratic power": "Separated from his impotentiality, deprived of the experience of what he cannot do, today's man believes himself capable of everything, and so repeats his jovial 'no problem' and his irresponsible 'I can do it,' precisely when he should instead realize that he has been consigned in unheard measure to forces and processes over which he has lost all control."[129] The implicit assumption here and elsewhere is that a fundamental impropriety is at work in the encounter of technology and life, which radically alters, in his view, the way human beings relate to one another and to themselves. The failure, however, to see that Agamben's superimposition of Heidegger's thanatopolitical reading of technology is at the heart of the reading of *dispositif* and democracy risks preventing us from noting the specific details about and differences in the contemporary moment in what Foucault called an "ontology of actuality."[130] The problem, if we choose to call it that, in Agamben's philosophy of life and its radical alteration at the moment that life interacts with technology is that death does not appear of its accord; rather, death is carried onto life's stage by technology. This suggests in turn that Agamben's analysis may come up short precisely when our focus moves beyond or outside of a Heideggerian critique of technology. It is as if Heidegger's critique of proper and improper writing risks introducing the thanatopolitical whenever technology is the subject and object of discussion. If we are to locate an ontology of the actual in the present, the problem with Agamben's reading of life is that it becomes difficult, if not impossible, to discover where the thanatopolitical drift enacted by a critique of technology ends and an effective ontology of the actual begins.

The thanatopolitics that seems to follow in the wake of appropriating an ontological critique of technology is a difficulty that Esposito avoids, but only in part. Certainly embracing the impersonal as a way of counteracting thanatopolitics on its own terrain brings risks as well. Above all, affirming the impersonal might well mean agreeing to a sort of resigned "everydayness," as Heidegger described it, which some, such as Agamben and, for different reasons, Badiou, will find abhorrent because it merely intensifies the impotence of the modern individual or collective (and hence the importance of avant-garde art as promoting the inhuman for Badiou). Another philosopher, however, has recently taken up less explicitly the relation of biology and politics. He, too, while elaborating a perspective on questions as broad as colonialism and the genome that is deeply indebted to Heidegger's critique of technology, has sketched a different reading of technology, specifically with regard to bioengineering. It is to Peter Sloterdijk's reading of Heidegger, technology, and immunized life that I now turn.

3 BARELY BREATHING

Sloterdijk's Immunitary Biopolitics

THIS CHAPTER grows out of an abiding appreciation of that philosopher who, along with Giorgio Agamben and Roberto Esposito, has attempted to think through the aporiae of the biopolitical. In those important pages in which the thanatopolitical springs forth, German philosopher Peter Sloterdijk explicitly deploys a Heideggerian perspective on developments in biotechnology as a way not only of justifying biotechnological "improvements" in humanity but indeed of offering a wide-ranging defense of biotechnology read in a medialogical key. These essays, in particular "Rules for the Human Zoo" and "Domestikation des Seins: Die Verdeutlichung der *Lichtung*," caused an enormous uproar in Germany at the time, given the warm Nietzschean overtones with which Sloterdijk appeared to paint Nazi eugenics.[1] My own impression, which I have sketched elsewhere, is that Sloterdijk's reading of Nazi zoopolitics was based on a notion of humanizing and bestializing media, of inhibiting and disinhibiting media, as he calls them, that failed to register the Nazi's biopolitical agenda outside of a media perspective.[2] Like so many other media theorists before him, Sloterdijk risks promoting a deterministic reading of technology in ways that hobbled both the work of Marshall McLuhan and also more recently that of his fellow countryman Friedrich Kittler.[3] Said differently, the media theory employed by Sloterdijk and authorized by Heideggerian ontology is much too reductive to explain the zoological politics of Nazism.

In the following pages, I want to take up again a reading of Sloterdijk in three more recent texts in which he continues to mine Heidegger's reading of technology in ways that recall both Agamben and Esposito. The result will be another crucial, if unsettling, perspective on the thanatopolitical.

The first concerns one of his important statements on political philosophy to date, the last chapter of the second volume of *Sphären*, titled "Die letzte Kugel." Here Sloterdijk adopts a Heideggerian perspective on globalization that skirts in and out of the thanatopolitical.[4] In the second, separation and its thanatopolitical effects are thought through a notion of environment and "atmo-sphere," which sets the scene for some of his most original and troubling insights in *Terror from the Air*.[5] In the final work, *Rage and Time*, thanatopolitics grows out of an impolitical perspective on rage, in a "modernization of ancient *menis*" and "transformation of the subject into an active gathering place of world rage."[6] All these writings share Sloterdijk's deep ambivalence about any future community, global or otherwise, and at the same time a tragic configuration of life cut off from community, one merely protected and secured thanks to modern technology whose effects are to subvert or attack a "proper being at home."[7] Sloterdijk's oscillation between proper community and improper households is where I want to mark the primary role of technology in his version of thanatopolitics, where the devastation of being is registered most fully in the many ways technology uproots and alters the proper abode of being today. I also point out on various occasions throughout the chapter where Sloterdijk himself fails to follow through on these reflections, especially when the subject turns to biotechnology and globalized gene pools.

One final note: by focusing exclusively on the thanatopolitical in Sloterdijk, I may appear to be drastically reducing the expanse of Sloterdijk's thought, which, given the volume of his output as well as his stunning expertise in areas running from cartography to European letters to linguistics, is substantial and deep. By focusing on the thanatopolitical and technology, in other words, I risk eliding a number of other possible rubrics with which to judge and evaluate Sloterdijk's thought. My response, the details of which will become clearer as I proceed, is that if we take Sloterdijk's perspective on globalization as our primary focus (authorized, in some sense, by Sloterdijk's equation of globalization and the notion of the sphere), it is impossible not to register a powerful inflection of globalization toward the thanatopolitical in his thought, indeed, to see globalization as a form of devastated being that ought to be called by its proper Foucauldian name of the thanatopolitical. This becomes clear in those important chapters Sloterdijk dedicates to globalization in the second volume of *Sphären*, to the central role it occupies in his more recent genealogy of breathing spheres in *Terror from the Air*, to the role of proper and improper forms of anger and rage and their associated technology in *Rage and Time*. All of which is to say that the thanatopolitical, incidentally never called as such by Sloterdijk, has moved to the center of his reflections and merits the position I want to award him in the pantheon of those adopting a thanatopolitical perspective on life and technology today.

The Space of Globalized Death

I want to begin by recapitulating Sloterdijk's perspective on globalization. Here the key term in his analysis, not surprisingly, will be space. Acting almost as if he were Heidegger's *doppelgänger*, Sloterdijk will see himself as bringing to completion Heidegger's uncompleted text that was to have been called *Being and Space*.[8] Thus, in a recent text, he writes, "The *Spheres* project can also be understood as an attempt to dig up Heidegger's project titled *Being and Space*, which remains unthematized in the first work of Heidegger (or at least not thematized in its fundamental features)."[9] For Sloterdijk, this missing treatment will take the form of ontology as the dimension before space, as the element that spatializes space—as a sort of matrix for dimensions in general. Sloterdijk will name this originary space "sphere," that is, as the realm wherein dimensions are disclosed, one that he will increasingly link to the notion of environment and climate as essential for life (in contrast to what he will describe as natural spaces). His mapping of globalization as a history of movement takes place through the relation between natural space and the human sphere.

Sloterdijk articulates the relation between these spaces across a staggering number of pages—the twists and turns of a universal history of space obviously merit attention. For my purposes, I propose rather that we take up those terms at the heart of the present study, namely, the one between proper and improper forms of writing and revelation, because it was in their oscillation that we first measured the thanatopolitical drift in Heidegger and then in Agamben. How does Sloterdijk theorize the relation between proper and improper? He does so in two related ways. In the first, he will read modern man in ways deeply similar to Heidegger's Western man of *Parmenides* as the figure (of life, of being) who comes to recognize his own distance from a transcendental outside. Thus, reflecting on Alexander Humboldt's representation of the world in *Kosmos*—though clearly other examples from *Sphären* could be cited—Sloterdijk will note the principal features of what he calls the final sphere. These he will link through a relation to modernity:

> The essential transcendence and the dream of a true fatherland in the beyond have irretrievably been lost for modern man [*Neuzeitmenschen*]. The self-reference of the subject thought as the condition for the return of that which is outside to what properly belongs to modern man emerges then.[10]

Following Heidegger's reading of Hölderlin's "Heimkunft/An die Verwandten," Sloterdijk will put forward the modern figure of *homo habitans,* who, in suffering the loss of transcendental and cosmic insurance, is made homeless from the transcendental fatherland. Such a space had been projected as a safe outside that suddenly has now become not only unreachable but profoundly indifferent to future existence. Why is the loss of protection offered by a transcendental homeland here understood as an outside that was formerly man's own? According to Sloterdijk, the loss of protection occurs when Humboldt, along with Martin Behaim, Johannes Schöner, Peter Apian, and others, mapping the cosmos, take up a second order of observation "on their planet."[11] This allows them to admit "that external spaces are nothing other than extensions of a uterine-social, domestic fantasy established regionally."[12] In other words, once the outside is no longer imagined as being the equivalent of a space domesticated by man (here we can sense the implicit war between the sexes that governs the move from transcendence to immanence in the passage), mankind falls back on the earth as that space that is properly its own. In the opening toward the infinite heralded by Humboldt, among others, man's sense of inside and outside is dramatically altered.

Consider, in this regard, the passage from Heidegger's essay on Hölderlin's poetry that I included earlier. Heidegger is speaking of the Hölderlin poem "Heimkunft" and the notion of "homeland":

> Suevien, the mother's voice, points toward the essence of the fatherland. It is in this nearness to the origin that the neighborhood to the most joyful is grounded. What is most characteristic of the homeland, what is best in it, consists solely in its being this nearness to the origin—and nothing else besides this. That is why in this homeland, too, faithfulness to the origin is inborn.[13]

In "Die letzte Kugel," the final chapter in the second volume of *Sphären,* Sloterdijk will read the fatherland outside of any reference to a national community, be it German or any other; instead, he moves the coordinates of the fatherland outward so that what qualifies as outside becomes simultaneously immense and unthinkable. What before was outside no longer looks like a community, or the "essence of the fatherland," in Heidegger's words, but rather takes on the features of something profoundly unknowable, a space that cannot be thought in terms of nearness or distance. The outside *(Das Außen)* "expands within itself, indifferent to the place of human beings, as an extraneous mass that is sovereign to itself. Its first and last principle seems to be that of having no concern for man."[14] Such a radical indifference on the part of the outside—the nihilistic and Nietzschean tones will not have gone unnoticed—does not mean that Sloterdijk will forgo the Heideggerian "nearness which still holds something back in reserve," that proximity that both brings near and distances.[15] The question for Sloterdijk will be how to inhabit the distance while never directly coming into contact with it—without even directly experiencing it. In ways quite different from a whole tradition of Italian weak thought and how to inhabit distance, such a distance for Sloterdijk cannot be inhabited properly as belonging to man.[16] Rather it is a nearness that is not to be thought as compressed or extended. Man experiences this distance "through his bodily and touristic extensions" *(mitsamt seinen Körperausweitungen und touristischen Extensionen)* and so becomes that modern figure of *homo habitans* who is really the improbable protagonist of so much of Sloterdijk's work.[17] Taken together, essentially, Sloterdijk argues that globalization is the making of the earth as man's home inhabited by the *homo habitans* and that it cannot begin at all until the outside has been marked as radically external to mankind. This move opens the way for the "discovery" of the earth through navigation, transforming it into a global sphere. Paradoxically, recognizing a threatening outside, one privileged by modernity, places humanity at risk. A proper space of the earth as home is fundamentally crossed by another space that is utterly indifferent to man. This indifferent and radically other space, working as a kind of nihilistic contamination, will soon be projected in European colonialism and totalitarianism onto the other outside spaces across the globe.[18] It is from those spaces that the European requires protection—a protection that will lead to employing death as a means to securing life.

Just who or what is this *homo habitans*? Perhaps a metaphor may help us see how Sloterdijk envisions him. The *homo habitans* is reminiscent of an astronaut—indeed, Sloterdijk will conclude the chapter with a reference to Neil Armstrong—who can only live outside his spacecraft with the help of a space suit that protects him from space by creating another interior space (a sphere) that allows him to experience this outside space as radically different from the domesticated space in which he breathes. The space suit functions as the means by which the individual moves through and acts within this radically other space, while remaining apart from it. For Sloterdijk, modernity essentially consists of the struggle to create these metaphorical space suits, immunitary regimes, he will call them, that will protect Europeans from dangerous and life-threatening contact with the outside (outside understood in the nineteenth and twentieth centuries as the imperial heart of darkness and as the ruinous effects of too close a proximity to one's neighbor in twentieth-century totalitarianisms).

This reading of the *homo habitans* as the immunized actor who moves through a dangerous outside will be immensely productive for Sloterdijk's thought, especially for his later readings of terrorism, for instance, as well as for his more literary-critical output.[19] Even more important will be how Sloterdijk uses these modes of protection to read the advanced stages of contemporary globalization and the threats it poses to humanity. To keep with our spacecraft image, the earth in this scenario now takes on the aspect of the space that is properly man's and as such can be explored, mapped, exploited, and destroyed when necessary.

What constantly returns in Sloterdijk's account, be it of the cosmos, of early navigation, or of colonial and postcolonial space, is the notion that mankind's "universal" task is to develop the means with which to interact with it, navigate and work in it, all the while remaining apart from it. This accounts for technology's decisively immunitarian inflection in Sloterdijk's thought. Immunity, of course, has a long and distinguished history in contemporary political thought, and as I have had occasion to discuss elsewhere, Roberto Esposito and Peter Sloterdijk's thinking of immunity has much in common.[20] For my purposes here, however, it may be more productive to highlight their respective differences so as to better sketch the topography of contemporary thanatopolitics. One of the most important differences, aside from Sloterdijk's deeply Heideggerian focus and Esposito's more vitalistic, Deleuzian one, is that Sloterdijk does not offer anything approaching the grand systematic treatment of community and immunity that one finds in Esposito. Instead, we get only fragments in which immunity is linked principally to an individual subject. In the following section, I want to consider the most suggestive and illuminating sections of his most recent attempts to think immunity—without losing the important context in which Sloterdijk offers them—as a way of furnishing a more complete philosophical perspective on the thanatopolitical. For Sloterdijk, the thanatopolitical cannot be thought apart from contemporary and individualized forms of immunity and the devastating effects they have on community.

How does immunity allow Sloterdijk to imagine the relation between this absolutely foreign and improper space (since it cannot belong to mankind, given its radical otherness) and the proper space of the earth? For Sloterdijk, this becomes possible only because "those human beings who are disposed to risk-taking . . . were not able to preserve or regenerate during the voyage and on fundamental endospheric conditions."[21] In other words, as modernity "progresses" (as globalization, but also via assorted crimes of the colonial variety, as Sloterdijk admits), it does so thanks not only to technologies of transportation but also to the capacity of technology to create spheres of individualized space, those endospheres that can be defended and renewed. This insight will lead Sloterdijk to the heart of his reading of globalization as the continued refinement of the practices of immunology: "For this reason the true history of terrestrial globalization must be narrated in the first instance as a history of spheres of protection that are brought inside and traversed by these salvific wrappings, be they visible or invisible."[22] There is much of interest here. Certainly the Heideggerian-inflected reading that he will soon offer of European colonialism as a decisive space in which "eco-technologies" are perfected or his emphasis on a worldly campground interests us as it depends on a level of protection, of proximity and distance between the proper and improper, that allows the European to perceive "the other through a theory-window."[23] The important result will be to make the outside livable, though we do well to note that implicit thanatopolitical figure that Sloterdijk elides here: the deaths of millions so as to make the colonial power safe far from "home," what Achille Mbembe in a not so different context will call "necro-power."[24] The question we need to ask is how these spheres of protection, with their inscription in a horizon of death, are to be thought today in the next phase of globalization, once capital has globalized the earth. When that occurs, these spheres take on an even more intense immunitarian function.

Sloterdijk places immunity at the heart of his theorizations of contemporary capitalism in *Sphären* in a section significantly titled "Das große Interieur." It is an apt description because in the narrative that Sloterdijk proposes, the contemporary moment is characterized principally by the construction and protection of increasingly fortified internal spaces, indeed, of a multiplicity of interiors that are premised on the centrality of individual life as the contemporary agent of protection. In some of the most compelling pages Sloterdijk has authored, he assaults the idea, put forward by Habermas and others, of a new postnational constellation that substitutes for modern political communities.[25] His is a devastating attack that begins and ends with a radicalization of the previous moments of ecotechnology, which is to say, their extension to the individual. The individual form of life today will be the one immunized:

> In this context [of globalization] the epochal tendency towards forms of individualistic life discloses its immunological meaning: in today's advanced "societies" it is individuals, perhaps for the first time in the history of the convergence among hominids, who, inasmuch as they are bearers of immunitarian competencies, break away from collective bodies (which they had until that time protected) and en masse now want to separate their own happiness and unhappiness from the preservation of the form of common politics. Today we are probably living the irreversible transformation of a collective politics addressed to the security of groups with an individualist immunitarian design.[26]

For Sloterdijk, we are living today (or perhaps dying today) the end of the political community as it was constellated during modernity to the degree that the individual forms of life function as agents that immunize biological life against forms of political life that are not individualized, namely, collective forms. Detaching themselves from the collective bodies to which they belong, they move toward other political collectives whose function is directed principally to individual security. Where previously, cosmic space had been externalized to such a degree that mankind was left without any transcendent moors, leading men and women to see earth as the proper space of home (and with it the associated forms linked to that collective space of protection, the national community, in particular), Sloterdijk now perceives a process in which ecotechnologies progressively become the domain of individuals. Today these individuals are charged with devising their own immunitarian forms of protection.

Today's individual who breaks away from the collective body follows directly in a line that moves through a number of different figures, including especially European navigation as well as architecture, which elaborated the form of a proper interior through the sphere.[27] Thus Sloterdijk writes in the second volume of *Sphären* with reference to forms of locomotion, particularly the ship: "The ship is a techno-spheric-magical self-expansion of the ship's crew, and in this it is, as is true for all modern vehicles, a homeostatic machine for producing dreams, that allows itself to be navigated like a manipulatable Great Mother through the outside element."[28] What distinguishes these homeostatic machines of yesterday from today's immunitarian designs is not only the telescoping of ecological technologies onto the body of the individual itself but the diminution of the insuring role, the protection that previously was provided by these collective bodies. Here Sloterdijk has is mind not only the national community or state but a general system that would include religions in what he calls "this indispensable transcendental system of insurance."[29] Today, with the end of transcendental insurance, what takes its place is nothing other than those individual schemes of protection, schemes immanent to she who holds them.

Sloterdijk will go on to theorize the likelihood of an immunological war whose purpose will be to artificially create conditions in which a lessened common bond may become the norm. Leaving that aside for the moment, what makes an impression here is the equation of security with individual forms of life. Sloterdijk's formulation emphasizes less the Hobbesian Leviathan of the sovereign or state who protects the body politic's members and instead evokes the intense pages of *Of Man*, in which Hobbes puts forward a notion of extraordinary power as good "because it is useful for protection, and protection provides security."[30] Sloterdijk, much like Hobbes, and Schmitt before him, also sees the increase in individual protection as good because the increase of prosthetics and locomotions (which turn inward toward the individual in the modern period in automobiles, trains, airplanes, tourism, etc.) creates the possibility for transforming the individual into a homeostatic machine who will merely look out for herself when faced with the Outside. In ways that both recall and differ from Agamben's *The Coming Community*, Sloterdijk will argue that power, here understood almost primarily as security, has shifted inextricably to individual as opposed to communal forms of life—to *zoē* as bare life apart from its communal counterpart, *bíos*. Sloterdijk will distinguish between communal and collective to the degree the former one-dimensionally addresses protection and security and little else. If Sloterdijk is to be believed, today we are witness to not just a dismantling of state protection that looks forward to some postnational figure of protection but rather a much more radical and troubling move toward securing individual life-forms through noncommunal entities.

Here we register the first drift toward the thanatopolitical in Sloterdijk's thought that occurs when communal life is continually exposed to pressure by private, individual immunity designs. The next step toward a full-blown thanatopolitics occurs soon after, when the death of some is used to protect the lives of others. We can see this in the assumptions Sloterdijk makes about the transformations that technology effects on man and community. To see how, recall those paragraphs I glossed earlier from Heidegger, when probing the effects of technology on how man's essence unfolds, Heidegger notes that asking whether man will be able to master technology is the wrong question—the right one is "what kind of man [*welche Art Menschen*] is alone capable of carrying out the 'mastery' of technology"?[31] Sloterdijk will take up Heidegger's emphasis on the individual man as the key operator and interlocutor of technology, who will then find in postmodernity the moment when technology masters man individually by creating seemingly ever more effective immunological containers. These will spell the ruin of political communal life. The reality for Sloterdijk is that contemporary forms of technology have reached such a stage of impropriety that new designs *(erfolgreich Designs)* will become available to those individuals who "are able to delineate with success liveable immunitarian conditions: and these conditions are those that can be developed and will develop in various modes with the societies of the permeable walls."[32] Such a possibility was implicit in Heidegger's reading of the hand and the relation among man beings and Being. Recall in particular those pages from *Parmenides* in which Heidegger links the notion of care to the hand:

> From this it is clear how the hand in its essence secures the reciprocal relation between "beings" and man. There is a "hand" only where beings as such appear in unconcealedness and man comports himself in a disclosing way toward beings. . . . The hand acts [*Die Hand handelt*]. The hand holds in its care the handling, the acting, the acted, and the manipulated [*das Handeln, das Gehandelte, und Behandelte*]. Where the essential is secured in an essential way, we therefore say it is "in good hands," even if handles and manipulations are not actually necessary.[33]

Sloterdijk emphasizes the metaphor of being in good hands as being properly secured and then does Heidegger one better. He reads technology not just as improper writing but also considers technology in the modern but especially the postmodern period as the design of blueprints for individual immunization that depend on nothing other than the improper use of the hand. The key term is *care,* but now it is a hand that no longer holds in its care all forms of action and, with them, the relation of man to others. Instead, the hand merely secures those individuals in lieu of proper care (of community, of the nation, of Hölderlin's homeland that has gone missing). In other words, in these individually designed immunity regimes, we have an improper securing of man's essence based on a form of technology now understood as schemes of private insurance that substitute for one's own national community: "This reveals a situation in which the majority is ever more able to dissociate itself from the destiny of its own political community . . . its immunological and national collective *optimum,* and not in systems of solidarity of the proper community."[34] The thanatopolitical is the name given to the end of communal protections and the death by exposure of those left without protection.

Let's also observe that Sloterdijk, in "Die letzte Kugel," will quite clearly associate this inscription of outside and inside with the phenomenon of neoliberalism. One of the principal effects of neoliberalism, he will say there, is that "within markets no one is properly 'at home'; no one needs to try to be at home there where money, goods and the fictional objects of the owner are exchanged from one owner to another."[35] The reason simply is that capital has transformed what was a house for all into a market for everyone: capitalism does the work of death through a world market that threatens humanity with an ecological disaster of catastrophic proportions—not measured merely in thousands killed by a reinforced biopolitical sovereignty but by the earth itself, which no longer is properly man's. This is the cause and effect for the emergence of immunitary regimes designed by individuals.

Yet it would be misleading to say that in Sloterdijk's judgment, the earth is everyone's and belongs to no one and that into that vacuum step those with enough money and licentiousness to build their own immunitary walls. In some of the most important pages of the second volume of *Sphären,* Sloterdijk, like Agamben and Esposito, will make reference to those lives that remarkably resemble *homo sacer* but that result from the processes of globalization:

> The unity of human beings within their own scattered species is today based on the fact that everyone, in the respective lands and histories, have

> become supplanted, synchronized, battered, and humiliated at a distance; grouped together and oppressed by excessive pretexts—simple sites of their vital illusion, addresses of capital, points in the homogenous space to which one returns and which turn on themselves—persons who do not see, but are seen; persons who do not understand, but are understood; who do not join, but are joined. Humanity after globalization is composed in the largest part by those who have remained behind in their own skin; of victims of the disadvantage brought on by the location of the "I."[36]

Sloterdijk assumes that to be a person is to be seen and addressed primarily as an object of capital because in this way, capitalism can depersonalize her, in a process that, I noted previously in my reading of Esposito, involves the stripping of the gift of grace from that part of the person that can receive it. Sloterdijk fills in some of the details of contemporary forms of person on which Esposito touches when discussing philosophical and religious fascination with personhood. Indeed, Sloterdijk suggests that there is something else going on: in the *dispositif* of person, neoliberalism finds its privileged mode of awarding charisma to that part of man that can dominate the animal part. Neoliberalism, through the "truth" of the market, becomes the final arbiter of who has made sufficient moves toward deanimalization. Two moves: those who are willing to do what it takes to get to care (to construct their own care through individual immunitary regimes) and those who do not—who are then addressed as less than persons, as remaining in their own prostheticized skin.

Thus another feature of contemporary perspectives on the thanatopolitical becomes visible. According to Sloterdijk, the space of globalization leads paradoxically to the privileging of the first person among those persons who are most directly addressed by capital. Here the reference to "vital illusion" indicates what the continued pressure of the outside that moves inside means for globalized humanity. A connected world, which appears to include everyone, has a complementary function of exclusion, a "super inclusionary system."[37] In a process we know well, while some are included in the increasing connectedness of the global village, becoming part of a first person plural, many others are excluded, limited merely to the first person singular. I use the grammatical term consciously, of course, because it rhymes with the notion of person that informs all of Sloterdijk's analysis. All of which is to say that where there is a *we,* there is also an *I,* but in a globalized world, this *I* is slotted as that part of humanity that is depersonalized. In other words, the *I* functions as a cover for the process of depersonalization.

The thanatopolitical for Sloterdijk is prefigured in the forced move from the first personal plural to the first person singular. Such a perspective differs from either Esposito's or Agamben's because Sloterdijk here sees depersonalization as working through (or moving back to) a sense of what is proper that resides only with one's body. *I* here would be emptied of all interiority and instead would mark simply the shell of the former *I* that was not excluded. That this occurs through an emphasis on the *I* and not the *we* should give us pause. What Sloterdijk appears to be saying is that those who are addressed as I no longer are afforded the protection of we, understood as those who see, those who understand, and those who reach out. Again, this does not mean that those *Is* that reach out to we find themselves the proper care that is guaranteed or offered by community, be it national or ethnic, finding it *bei sich*. Instead, they, too, are thrown back on themselves, with the implicit understanding that in the accelerated circulation of objects and persons as objects, they, too, may find themselves pushed back to the first person singular from the plural.

The Terror of Immunity

In a more recent text, Sloterdijk has elaborated the features of individual designs of immunity that can help us pinpoint the thanatopolitical coordinates of his thinking to include the nature of breathing itself. In *Terror from the Air,* he extends his analysis of the individual beyond his body to include what are essentially individual spheres—individual atmo-spheres (where *atmosphere*'s etymology is "breathing sphere"). In a chapter titled "Air/Condition," Sloterdijk will offer a biopolitical reading of the twentieth-century avant-garde in a context of heightened fear and demands for individual protection. Introducing a 1936 Hermann Broch speech on breath as his starting point, Sloterdijk extends his notion of the individual subject to the space of breathing and with it the modes by which breathing may be repressed:

> In this characterization, Broch's narrative art is held to rest upon the discovery of atmospheric multiplicities, through which the modern novel was able to go beyond mere representations of individual destinies. Its subject is no longer individuals in their entangled actions and experiences, but rather the extended entity of individual and breathing space. Its plots no

> longer take place between people, but between respiratory economies and their respective residents. The ecological vision permits us to place modernity's critical motif, that of alienation, on different foundations: it is the atmospheric separateness between people that assures their embedding in "economies" specific to each of them; the unavailability of some persons to differently-attuned, differently-enveloped, and differently air-conditioned others is indeed more and more evident. The breaking up of the social world into spaces of moral independence inaccessible to one another is analogous to the micro-climatic "fragmenting of the atmosphere" (which in turn is correlative to the fragmentation of the "world of values").[38]

Sloterdijk suggests that a dramatic shift has taken place in the relation between the individual and her atmosphere. A shared space in which breathing could take place among individuals—what Sloterdijk calls the ether—cedes ground to individualized atmospheres in which shared breathing across space is blocked in favor of individual households. More important though are the effects of the breakdown on the unity of the individual and breathing spaces. Exchanges (though perhaps we might want, following Heidegger, to speak of actions, or *Handlungen*) between individuals are no longer carried out between "persons" but breathing households and those who inhabit them. The implication is that the condition of persons who act is no longer available when households assume the place formerly occupied by a shared breathing space. Indeed, if we were to follow further this line of thought, we would sense that Sloterdijk assumes that the notion of person changes when agency is awarded to these "breathing households."

Here we have another take on depersonalization that, pace Esposito, moves through immunity. Sloterdijk sees in the immunization of spaces and breathing the modern process whereby power no longer addresses subjects as "individuals in their entangled actions and experiences," and no longer as persons either (in the same way that Broch's plots "no longer take place between people"). Rather power is directed to individuals in the space of their breathing. These households, we should be clear, are alive as well as armed with agency, which makes sense because their job is to secure the limited and separated breathing space of their own respective households. The result, though Sloterdijk never states it as such, is a modification not only of the nature of person as a concept but also of the individual and community. There is not, as one finds in Esposito, the individual on one side and the community on the other (or in this case, shared breathing spaces) and their reciprocal generation of immunity. Instead, we have an individual who, to breathe, must be separated from other individuals through the *dispositif* of the household. The household will be the improper entity within which those with similar immunization protocols will be housed. In lieu of persons to whom the market awards grace so as to deanimalize them, Sloterdijk thinks that the notion of person will give way to another entity that will only be able to breathe the same air as those others from similarly immunized spaces. For Sloterdijk, immunization today leaves behind the mere immunization of the body that, for postmodernism, was what allowed for the extremely high tolerance the self enjoyed when faced with the other.[39] Immunization is now to be understood ecologically. Yet let's be clear: the entity who inhabits the household's microclimate has been depersonalized, not toward the animal (Sloterdijk is mum here about the features of these inhabitants) but rather toward some living thing that is much less active or reactive.[40] Once the category of person has been sufficiently weakened by individual security ventures, contemporary immunity regimes will be centered on the household as it is capable of protecting breath (and hence mere life itself).

As I noted earlier in my use of *dispositif* to describe the features of the household for Sloterdijk, we are firmly in the realm of a biopolitical *dispensatio,* one in which the separation of individual households is what allows them their "embedding in 'economies' specific to each of them" and what permits them to be superimposed over other similar spaces so that they may interact. In this verticalization of economies of scale between similarly immunized atmospheres, interaction with spaces on the horizontal axis is impossible, given the undifferentiated space of what is outside the household and the impossibility of exchanging with them, given the control that each individual household gives itself over its own improper atmosphere. In this *dispositif* of the household, separation according to breathing regimes and air quality control gives rise to microzones of immunity, that is, living, breathing entities with little contact with who or what does not breath the same air. If I may be allowed the comparison, it is as if the former blood ties of family or the relation enjoyed with one's own birthplace had been turned outward from one's person to now include the breathing space of those whose individual immunitary designs most closely match one's own.

How might we translate this move to the immunized and immunizing household in terms of *bíos* and *zoē*? In *Terror from the Air,* atmospheric separation spells the creation of *zoē* to the degree that the household becomes the mode by which "the simple fact of living common" to all human beings is secured through their breathing. As separation into autonomous breathing containers or spaces continues, it becomes increasingly difficult to locate, let alone imagine, *bíos*: not an identity per se, but a living and a breathing that belongs to a political form of life. That, I would argue, is the threat posed, as Sloterdijk sees it, by individual households to humanity: that the separation is so complete as to make connections with other forms of life difficult. Yet I would go further. The creation of lives that breathe is not simply the final result of the processes, or more menacingly, procedures, that fall under the *dispositif* of the household. Instead, we ought to link it to liberalism, and especially neoliberalism. The thanatopolitical drift of Sloterdijk's argument is clear enough—the extension of individual power to the household precludes a political form of life that involves a shared breathing and, with it, a properly shared climate and, via metonymy, a shared *Weltanschauung*. *Bíos,* in the modern immunity of individual households imagined by Hermann Broch and elaborated by Sloterdijk, withdraws in favor of *zoē* and the barest fact of living common to living beings: breath itself. In this tragic perspective, a collective with bare living and bare breathing is fast becoming the norm.

Zoē as Bare, Breathing Life

The next question certainly has to be, when did we begin barely breathing? For Sloterdijk, that genealogy can be located first in the chemical attacks that characterized some of the early fighting on the western front during the First World War. Bare breathing becomes visible, so to speak, as a modern phenomenon first at the battle of Ypes on April 28, 1915, when the Germans launched their chemical attack through the use of toxic clouds. A number of elements stand out in these first chemical, terror attacks that allow Sloterdijk to link them to biological terror today. First, the attack was directed on the soldier's individual environment such that the body itself became one of the means for death:

> The attack on humans in gas warfare is about integrating the most fundamental strata of the biological conditions for life into the attack. . . . Not only is it the case that, as Jean-Paul Sartre remarked, desperation is man's attack against himself; more: the gas terrorists' assault on the air induces

desperation in those attacked, who, unable to refrain from breathing, are forced to participate in the obliteration of their own life.[41]

Terror, then and today, as he says, works by transforming man into the instrument of his own destruction. Moreover, we find the genetic feature of terror (as opposed classically in Hobbes to fear, which is always productive) working "against the very human-ambient 'things' without which people cannot remain people."[42] For Sloterdijk, terror is a technical form of violence or a violent form of technology that depersonalizes to the degree that it attacks the very air that a person breathes and in which she lives. As we saw in the preceding chapter when discussing Esposito's reading of the *dispositif* of the person, the sliding scale toward the semiperson to the nonperson is flexible enough that such attacks on individual atmospheres may have the effect of shifting large numbers toward the realm of the antiperson (attempts at Guantánamo Bay at psychological warfare through loud music are a case in point among many others).[43]

The thanatopolitics of terror becomes clear when we see the role that technology plays in making possible the destruction of environments. In the various technologies that fall under the title of climate control and meteorology, Sloterdijk sees them less as the effects of wartime technology or, for that matter, peaceful forms of knowledge and more as "sediments of war and post-war science."[44] In the resemblance between meteorology and terrorism, we glimpse the thanatopolitical because this form of impolitical knowledge allows its "user" to attack the conditions of existence of those who inhabit a certain space. "New weapons of terror are those through which the basic means of survival are made more explicit; new categories of attack are those which expose—in the mode of a bad surprise—new surfaces of vulnerability."[45] Again, we can make out the figure of Heidegger, whose writings on the question of technology are implicitly raised and answered by Sloterdijk, especially in the central term he uses throughout *Terror from the Air,* as "a form of exploration of the environment from the perspective of its destructibility."[46] In Sloterdijk's reading, terrorism exploits a modern technology of space exploration so as to be able to deny the very conditions of life, indeed, the primary conditions of life itself. Parallels with Heidegger's perspective on the protocols of improper writing come to mind as well, but they seem to have been made less metaphysical in Sloterdijk's account. The technology of the environment for Sloterdijk begins, as we noted earlier, primarily at the beginning of the nineteenth century, and its transformation of that space as one for manipulation and, implicitly, attack occurs during the First World War. The threat posed to mankind today is thus of a truly global sort. Yet, whereas for Heidegger, the threat was never spelled out in anything other than ontological terms of a withdrawal of Being from which man had to be saved, for Sloterdijk, the catastrophe has already arrived in terms of massive vulnerability caused by these forms of ecotechnologies. Man's very breath is put at risk. Yet something more troubling is hinted at as well, namely, that when the connectedness of breathing is broken across communities, be they global or local, the result is that implicitly, when some individuals try to breathe, others do not.

Here again we see how Sloterdijk inflects immunity thanatopolitically in the sense that immunity in the modern period is thought as one of defense against threats and less immunity as privilege. The birth of environmental terrorism—and terrorism is always ecologically directed for Sloterdijk—is seen as a response to these immunitary regimes that grow more sophisticated over the course of the last century. As globalization proceeds, more powerful immunitary regimes develop, in turn designed by individuals, often in joint ventures. Yet these designs demand in equal measure a form of technology that will challenge them by threatening to overtake defenses. Thus

> it is crucial to insist on identifying terrorism as a child of modernity, insofar as its exact definition was forged only after the principle of attacking an organism's, or a life-form's, environment and immune defenses was shown in its perfect technical explication.[47]

Sloterdijk's premise is an accelerating point–counterpoint in which terrorism breeds ever more powerful immunitary regimes that are able to deny to the terrorist access to the environmental conditions without which the human organism cannot live. Some must die, paradoxically, so that others in the future will live. The understanding, however, is that the game of one-upsmanship between them never ends.

We will want to raise objections with Sloterdijk's analysis, some of which I will return to in my conclusion. For now, I note only the implicit technological determinism that he assumes throughout his reading, in which immunology gives birth to modern terrorism. Similarly, Sloterdijk fails to see immunity as anything other than defense or to consider increasing immunity regimes (and heightened terrorist attacks, which they implicitly call forth in his analysis) as leading to what Derrida and others refer to as autoimmunity: the attack by immunity on the entity it was intended to protect (the individual or the body politic).[48] With that said, the nexus between technology and immunity protocols allows us to make out yet again the features of the thanatopolitical in Sloterdijk's thought from another perspective. He implicitly recognizes a fundamental truth of immunization, one that Esposito captures well in *Communitas* as the fundamental question of community: "How are we to fight the immunization of life without making it do death's work?"[49] The more one immunizes life, the more one calls forth death, and given that immunization accelerates in the modern and, especially, the postmodern periods, the form that this death will take will be increasingly ecological and global. It will also be communal in the sense that I noted earlier: in the place of communities, be they national or ethnic, we now have individual immunitary designs, which, in Sloterdijk's account, will become ever more protective and hence more vocal in calling forth more attacks. Said differently, rather than being merely the bad boy of German contemporary philosophy, Sloterdijk does in fact share a principal concern with many other European intellectuals writing today, and that concern centers around a missing community. The thanatopolitical drift that we noted in Agamben and Esposito as emerging out of an anxiety of technology with reference to the community is also to be found in Sloterdijk. For him, too, the thanatopolitical will be expressed in terms of the end of a proper community.

The Inhumanity of the Modern Avant-garde

This is not Sloterdijk's only perspective on immunity. He is not merely content to offer a tragic view of the irresolvable dialectic between more powerful immunitary designs and ecoterrorist attacks. Before I look at his other responses, however, it is important to single out another feature of Sloterdijk's thinking of these joint ventures in protection. That feature involves his impolitical reading of the modern avant-garde as practitioners of a certain form of terror that will, not surprisingly, have thanatopolitical effects on community. Sloterdijk does so in *Terror from the Air* by implicitly invoking the high modern view of innovation in the Imagist dictum "Make It New!"[50] Yet this innovation does not concern primarily ready-mades or question of form so much as it does the creation of a new society heralded by all the avant-gardes in which society becomes both the subject and object of "permanent revolution":

> Permanent "revolution" demands permanent horror. It presumes a society that continually proves anew to be horrifiable and revisable. The art of the new is steeped in the thrill of the latest novelty, because it emerges by mimicking terror and in a parallelism with war—often without being able to say whether it is declaring war on the war of societies or waging war on its own account. The artist is constantly faced with the decision of whether to advance as a saver of differences or a warlord of innovation against the general public.[51]

He will go on to speak of the modern masses integrated into "emergency communistic units," regimes of permanent fear, as he calls them, based solely on a common, threatened state shared by all. Again, so much of Sloterdijk's account of immunity here is implied that we need to proceed carefully, but obviously he makes fear and revolution coterminus, where fear emerges at the same time as calls for revolution both in the arts and in politics. In other words, catchphrases like "Make It New!" contribute to regimes of fear in dismantling forms of community in favor of a society that can be more easily frightened and controlled. These "emergency communistic units," which remain unidentified here, will elsewhere be referred to as totalitarian.

In such a scenario, a skeletal form of community appears just as the emergency is declared, providing a makeshift identity that is manifested only because of an (outside) threat. Sloterdijk does not qualify the nature of this identity, but it does not require much effort to see an implicit opposition here between a prior community that provides communal bonds and a qualified life or *bíos* and a present community under conditions of emergency that offers merely identity and hence weakened immunity protection. Without wishing to push his thinking into any kind of thanatopolitical straitjacket, Sloterdijk appears to suppose that attempts to make art new by the avant-garde at the opening of the last century weaken a politically qualified life to the degree that they mimic the terror of the new. The result is a turning away from community toward something else that is based principally on the fear of not having a community or being a community—an early version of identity politics that appears as the emergency, created and intensified as innovation, spreads.

The role afforded the avant-garde in this monumental move toward another form of emergency community (and emergency immunity) is immense because most of the avant-garde Sloterdijk examines will in fact advance themselves as "warlords of innovation." To understand fully the stakes of Sloterdijk's perspective, let's compare this perspective to that of another contemporary philosopher writing on much the same theme:

> Infinite in its act, art is not the least bit destined to the satisfaction of human animals in their tepid everyday life. Instead, it aims at forcing a thinking to declare, in its area of concern, the *state of exception*. Whilst accounting for the act, the qualitative infinite is what exceeds every result, every objective repetition, all the "normal" subjective states. Art is not the expression of ordinary humanity, of that which within the human obstinately strives to survive, or, as Spinoza would put it, to "persevere in being." Art bears witness to the inhuman within the human. Its aim is nothing short of compelling humanity to some excess with regard to itself. . . . In this respect, the art of the century, just like its politics or its scientific formalisms, is starkly anti-humanist.[52]

Here Alain Badiou posits that art operates in the state of exception and thus shows how mere *zoē* can, thanks to art, cross the threshold to *bíos*. For Badiou, art does not satisfy the human animal; it is not only directed toward reinforcing a bare form of life but rather pushes thought itself to consider art's relation to forms of political life, to *bíos*. What goes by the name here of the qualified infinite—the assonance of qualified with a qualified form of life called *bíos* is not happenstance—is what exceeds the act itself; not just any act, but all acts, and especially those objective repetitions and subjective states that go under the name of "normal." Consequently, an abyss opens underneath mankind. On one side, we have ordinary humanity engaged in repetitions linked to the ordinary and the normalizing, and on the other, an inhumanity folded into the human that art discloses—on one side, a form of ordinary life that merely survives (and the suggestion is that it won't survive for long) and another that does not merely survive but that lives on infinitely, an infinite qualified life or *bíos,* and hence not ordinary humanity but an exceptional inhumanity. On one side, we have a humanity not understood in its animality, the human animal, as what merely lives, and on the other side, the inhuman human purged of its animal nature.

In this scheme, art and the artist's function are to force humanity to come face-to-face with its own unqualified limits—the mere repetitions that characterize ordinary life, the normalizing therapies that create ordinary states while blocking an opening to the inhumanly infinite and infinitely inhuman. The differences with Sloterdijk couldn't be greater. Where Badiou sees the artist pushing humanity beyond itself, Sloterdijk sees instead a permanent performance of shock and awe dedicated to creating fear; by mimicking terror, the avant-garde artist helps dismantle forms of qualified life linked to community. Where Badiou sees *bíos* only emerging in the process by which art makes ordinary humanity aware of its normalized status—of which terror or terrorizing may well be a part—Sloterdijk sees art waging war on behalf not of some inhuman future but rather on behalf of fear itself. It is as if innovation, be it technological, political, or artistic, works as a multiplier of fear in Sloterdijk's analysis. So, too, do Sloterdijk and Badiou differ on the inscription of biopolitics ultimately into the horizon of community. Community is, of course, missing from Badiou's analysis. This lack of reference to community is due to Badiou's perspective, in which ordinary humanity is viewed as consisting solely of *zoē* while pushing *bíos* to some future transformation set in motion by an art that exceeds the mere repetitive act that is typical of bare life. Badiou translates the classic Marxian term *alienation* as fundamentally constitutive of *zoē*.

This reading of Badiou allows us to see once again where Sloterdijk's thought drifts toward the thanatopolitical. For Sloterdijk, community today is only a skeleton of its previous self, having been weakened repeatedly by modernity, but especially by those first terrorist forms of twentieth-century modernity that were called the avant-garde. In its place appear two forms of immunity: the individual and the emergency collective. Neither is based, however, on qualified life or the giving of the *munus* but rather on either the barest of immunity protections or the most advanced (and hence doomed to attack sooner or later). In this sense, Sloterdijk's thanatopolitics emerges out of qualified life withdrawing at the beginning of the twentieth century due in no small part to the appropriation on the part of the avant-garde of innovation and, with it, the possibility of intensifying fear across large populations. Through a series of layered semantic lexicons, the innovation practiced will be deeply inscribed in forms of technology that move mankind closer (and back) to a Hobbesian state of nature, in which individual immunity regimes attempt to take the place of the Leviathan.

The Thanatopolitics of Rage

In Sloterdijk's most recent work in English, these reflections on the biopolitical effects of the avant-garde are part of a larger project in which he traces the history of rage over the course of Western civilization. One may well ask what rage has to do with biopolitics. Clearly by referring to it and its close cousin, resentment, Sloterdijk is up to some of his old tricks, all pulled out of a Nietzschean hat. Leaving aside for the moment the Nietzschean "debt," a number of vectors appear here that pick up the threads of Sloterdijk's previous work, while others provide further details on thanatopolitical and technology. First, there is, of course, the fully articulated critical assault against Marxism, which Sloterdijk sees as practicing in its own obviously resentful way a sort of thanatopolitics by assuming a sacrificial economy when it comes to weighing the results of revolt. Here Heidegger's distinction between proper and improper writing and revelation will once again be translated, but now across proper and improper forms of rage. Thus Sloterdijk will link pure rage to the Greek heroes who were emptied of their interiority and their egos in almost impersonal fashion. Thus, Sloterdijk writes of Achilles,

> In the case of pure rage there is no complex inner life, no hidden psychic world, no private secret through which the hero would become understandable to other human beings. Rather, the basic principle is that the inner life of the actor should become wholly deed and, if possible, wholly song.[53]

The point is not that one fights for something but that the battle serves to bring forth a heroic fighting spirit or energy. The emphasis that Sloterdijk places on *Handlung* (action) in relation to pure rage inscribes this sequence in the horizon of proper and improper forms of technology as well because rage for Sloterdijk, in its production and transformation of events, will call forth energy. This was essential, we recall, for Heidegger's reading of technology and the protocols of improper writing in "Heimkunft."

Two hundred pages later, Sloterdijk traces the improper and exquisitely modern forms of rage and resentment to the French Revolution and, in particular, to those masters of rage, the Jacobins, reaching its maximum point of intensity with Marx and Lenin. What stands out in *Rage and Time,* aside from the broad brushstrokes Sloterdijk uses to paint what is to say at a minimum a complex and multidimensional history of Marxist thought, is the role of rage in the construction of the new man—the new revolutionary subject and the kinds of diagonal cuts that will separate this future vision of humanity from its decadent bourgeois predecessor. Interestingly, Sloterdijk will read the political avant-garde, in particular, the political party, be it socialist or communist, as functioning essentially as a sort of technology—as giving form to rage in a sort of dromological movement that Paul Virilio described in a not so different context.[54] Speaking of the semantic chain whose first link is the party militant, Sloterdijk will note that "modern militancy looks back on a long succession of rage corporations in the shape of secret societies, terrorist organizations, revolutionary cells, national and supranational organizations, workers' parties, unions of all shades, aid organizations, and artistic associations. All of these are organized according to conditions of membership, rituals, and club activities, as well as their newspapers, journals, and editorial houses."[55] The point of these bodies of rage, though clearly they resemble media more than anything else, is to push a people toward "subversive energy," to force them to give vent to their "explosive unhappiness."[56] Unlike their artistic brethren, these political avant-gardes harvest rage by giving it form. They provide it with a collective body with which and through which rage is embodied materially and collectively.

Sloterdijk will go on to associate the thanatopolitical with these improper forms, in particular, with the political party. We have in some sense here the other side of Heidegger's fear about the technologization of man under the Bolsheviks, the standardization wrought by the typewriter that slotted Western man into points in communication and lessened the individual features of his physical as well as written expression. Thus Sloterdijk translates this into a proper individual rage that in some respects seems almost impersonal as he describes it early on, compared with the massification or "The Thymotization of the Proletariat," in which the individual's expression of ire will be subsumed and multiplied in a systematic understanding of "alienation" and "reappropriation."[57] Sloterdijk goes on at great length to demonstrate how these terms contribute to collective rage, but the point of his reflections is to allow him to argue that the Marxist project will be fought over the terms of who or what is human and who or what is not:

> Nothing less was at stake with regard to the liberation of the working class than the regeneration of the human being. This liberation could correct the deformations resulting from the living conditions of the majorities in class societies.[58]

Despite capitalism's dehumanization of the proletariat via reification, the seeds for a future humanity reside there; it is "the true matrix of humanity in general, with all its future potential."[59] The enemies of the proletariat therefore are enemies of humanity and, according to Sloterdijk's reading, "[deserve] to be pushed back into the past."[60] Thus the thanatopolitical vector of Sloterdijk's analysis does not lie in agreeing with Marx's reading of the dehumanization of the proletariat (which, in fact, he does) but rather in the construction of the enemies of the revolution, these enemies of a potential humanity who must be pushed back into the past in this "ultimate war."[61]

Sloterdijk's language in describing the ultimate class and civil war envisaged by Marxism is also thanatopolitically derived:

> The ultimate war was supposed to release unconditional hostility: the capital-owing bourgeoisie, including its well-fed entourage, as the objective brutes, on the one side, and the proletarians, who were the sole producers of value, together with their escort of hungry offspring as the objectively true human beings, on the other side. In this war, the stakes include the true nature of the producing human being. Because one party, it is said, entertains a merely parasitical relationship with regard to production, while the other party includes those that produce authentically, the latter has to be rightly and inevitably victorious in the end. From this moment onward, to understand the essence of reality meant to think civil war.[62]

Sloterdijk clearly intends, by employing the language of extermination, to evoke the thanatopolitical universe of Nazism; so, too, does the distinction between real humans and nonhumans recall, if not Nazi biopolitics, then certainly a form of early twentieth-century anthropology and race thinking, all framed again in the language of proper and improper (authentic production), a distinction that Sloterdijk will read critically as the basis for the ultimate war. Obviously, there are significant difficulties with Sloterdijk's rhetoric here, but let's recall again that Foucault himself, in his seminar from 1975–76 at the Collège de France, collected in "*Society Must Be Defended,*" expressed himself in not so different terms from Sloterdijk when discussing the racism of socialism: "In addition to the State racism that developed in the conditions I have been telling you about, a social-racism also came into being, and it did not wait for the formation of socialist States before making its appearance. Socialism was racism from the outset, even in the nineteenth century."[63] Foucault will add shortly thereafter that "whenever, on the other hand, socialism has been forced to stress the problem of struggle, the struggle against the enemy, of the elimination of the enemy within capitalist society, racism does raise its head, because it is the only way in which socialist thought, which is after all very much bound up with the themes of biopower, can rationalize the murder of its enemies."[64] Essentially, Sloterdijk takes over Foucault's analysis of racism and reads it through the paradigm of rage. What both Sloterdijk and Foucault share is the recognition that in the thymotic revolution, a form of thanatopolitics evolves in which the potential human, carried by the proletariat as possibility, can only be actualized through a rage that justifies the murder of the bourgeoisie as less than human. They diverge in Sloterdijk's broader brushstrokes with which he paints Marxism and by forgoing some of the niceties of Foucault's discussion about social racism, since he translates it into wrath and rage.

Sloterdijk will supplement this reading of rage (or better, his reading of how rage was domesticated after the Greeks into its debased forms) by interpreting rage through the figure of the "misanthropic International," which reads like a crisis brought on by the effects of too much rage. Thus we find ourselves facing "an epidemic of negativity" that works from within civilization to produce revolts. Of course, the problems with such a perspective are numerous, but for Sloterdijk, most decisive are the features of this misanthropy, articulated not just by communism but by most, if not all, liberal forms [*des neo-kapitalistischen way of life*] as well. For Sloterdijk, the threat of the common is exactly that its purveyors only perceive the common as a massing, which ignores a fundamental truth about human socialization. The truth of the matter, according to Sloterdijk, is that humanity contains within it a strong tendency to sociophobia and that any attempt to institute the common by a party or a State does the work of the sadist. This is the ultimate horizon of the camp for Sloterdijk, and not surprisingly, it is one that will elide differences between Nazi, Soviet, or Chinese Communism. He writes,

> Rather, the camp rests on the intuition that hell is always other people as soon as they are mutually forced into unwelcome proximity. In *Huis clos*, Sartre merely replaces macro-hell with micro-hell. If one shoves one's enemies together into a state of total coexistence, one is responsible for each individual being burned in the small flame of induced hostility against his kin. Only saints survive camp situations without being dehumanized. "Camp" is only a conventional name for the modern forges of misanthropy.[65]

Thus bare life resulting from the state of exception is less important for Sloterdijk than the effects of rage—caused by a misanthropy brought on by the unbearable nearness to one's neighbor. If there is a process of dehumanization that takes place in these camps, for Sloterdijk, it is one caused by this toxic proximity to others.

The analysis is deeply unsettling for many reasons, not the least of which is its failure to take account of the specificity of the Nazi death camps and the importance of Nazi biocratic practices. Clearly we will want to raise questions about these claims. With that said, it is also clear that Sloterdijk sees a kind of original biopolitical (and metaphysical) mechanism that goes by the name of the thanatopolitical: we sense it in the opposition between forcible common-ization, on one hand, and the sociophobia of the individual, on the other. When he moves to a discussion of contemporary forms of anger and rage, he will, in ways similar to Agamben, also assume bare life as the minimum common denominator linking the masses in the West. Where they differ is what each chiefly sees as the features of bare life, which is to ask what precisely bare life lacks. For Sloterdijk, bare life lacks proper forms of rage; it is incapable of any "initiatory power to redeem one's own value and claims."[66]

The reason concerns the power of technology. Perhaps it no longer bears repeating, but this reappearance of proper and improper rage calls to mind again Heidegger's reading of technology. This allows Sloterdijk to bear witness to what is most properly individual—rage that has nothing of the impurities of resentment that Sloterdijk sees operating on the left because it does not install improper forms through the massification of its members. Today the catastrophe for Sloterdijk is not that man or individuals rage—that is not the point—but rather that this rage is transmitted in a society of the spectacle whose effect is exactly to subjugate all themes to the "law of standardization": "It is their democratic mission to create indifference by eradicating difference between major and minor matters."[67] The catastrophe will be the Heideggerian one we know so well of Western man, who now must be saved, but cannot, because of the power of mass media to create indifference. The dehumanizing of human beings results, brought on by disinhibiting media.

Another catastrophe awaits, however, and it rhymes with *What Is an Apparatus?* Sloterdijk will name the threat if Agamben does not: it is Islam. Writing in a tone that recalls the vitriol with which Oriana Fallaci attacked Islam, Sloterdijk will admit that even if "so far Islamism has little to show that would enable it to creatively continue the technological, economic, and scientific conditions of existence for humankind during the twenty-first century," the risk is that it soon will.[68] Combined with a demographic explosion within Islam, the catastrophe for the West is obvious enough for Sloterdijk—a biopolitical attack awaits. "It will not be an empty time of waiting for the West."[69] Again one finds many troubling moments in Sloterdijk's analysis, not the least of which is the monolithic view he provides of Islam (even if he does allow for the possibility of a cosmopolitan Islam to emerge later on): the Manichaeanism at work here in fact detracts from Sloterdijk's contributions to our understanding of the contemporary biopolitical moment. With that said, we can see that Sloterdijk has essentially substituted Islam for communism in Heidegger's analysis of proper and improper writing. There we recall that Heidegger spoke of the technologized Bolshevik and "the complete technical organization of the world."[70] Sloterdijk, on this score, hedges his bets. On one hand, the coming catastrophe is that Islam, armed with technical know-how and given the viral epidemic of indifference in the West, may well gain the upper hand. On the other hand, Islam lacks the sheer anger, the banks of rage on which communism cashed its checks. Sloterdijk, in other words, will have none of the Christian, Islamic, or Communist traffic in apocalypse. The reason is unambiguous enough. The catastrophe is already long past us, thanks to globalization and the immunity regimes that characterize it. In such a scenario of postcatastrophe, biotechnology, with its domestication of being as well as rage, is the only strategy that retains any power to alter radically the rampant dehumanization and the surrounding indifference that characterize Western life.

The Thanatopolitics of Biotechnology (I): Foucault and Sloterdijk

In the pages that remain, I want to tie up some of the loose ends of my discussion of Sloterdijk's thanatopolitics, which begins with immunity and ends impolitically with a defense of heroic rage as one of the few avenues available for life to dezoologize itself. Sloterdijk's discussion raises a number of points of contact with areas generally thought off limits in his perspective on globalization. In some ways, for instance, his analysis rhymes with Italian *operaismo* (workerist philosophy) and immaterial labor and perhaps, to Sloterdijk's horror, has points of contact with Toni Negri as well.[71] Be that as it may, let's observe as well that the drift toward the thanatopolitical does not occur all at once in Sloterdijk's model of sphere, globalization, and improper forms of anger—he did not just wake up in 2007 to discover that proper care begins at a nonexistent home and that mankind must continually fight for it. The process occurs in history, and so Sloterdijk is not merely following an antihistoricist line of thought, say, as in Heidegger's *Parmenides*, given the sheer detail he offers on the genealogy of globalization. We recall how he traces the Outside's increasing power and the concomitant response of the inside to construct virtual individual walls.[72] Rather, as I have attempted to show here, a large part of his thanatopolitics really concerns the death of community, which explains the tragic tonality with which Sloterdijk treats it in the second volume of *Sphären*. Thus globalization for Sloterdijk is not empire; there can be no "super monosphere or a powerful center of all the centers," no virtual or rhizomatic empire that could provide a newly charged notion of the common, but only a future of ever shifting joint individual ventures that involve different groups who share the same interests at any given point in time.[73] Whereas before, family or community provided proper care, that possibility no longer exists. Thus the thanatopolitical catastrophe has less to do with Western man or humanity in general than it does with the death of community and the rise of individual immunity. In lieu of community, contemporary societies composed of individual members now gain momentum. The essential strategy they adopt, indeed, one of the only strategies available, is to design ever more powerful immunitary regimes based on flexible networks of interest.

The relation of immunity and community calls to mind not only Esposito's understanding of the term in relation to community—and here many points of contact with Sloterdijk are already evident. Even more central is Foucault, as he should be, for a discussion of globalization as individual biopolitical regimes that subvert communities. In the lecture of March 17, 1976, collected in *"Society Must Be Defended,"* Foucault treats the environment in quick fashion in his genealogy of biopolitics. What he says echoes, however, Sloterdijk's reading of ecology and the end of community:

> Biopolitics' last domain, is finally . . . control over relations between the human race, or human beings insofar as they are a species, insofar as they are living beings, and their environment, the milieu in which they live . . . and also the problem of the environment to the extent that it is not a natural environment, that is has been created by the population and therefore has effects on the population.[74]

Recent reflections on Foucault's elaboration of biopolitics often elide this ecological perspective, but it is one that Sloterdijk, in some sense, returns to and overturns. Where Sloterdijk pushes beyond Foucault is in having deconstructed populations as representing a moment prior to the current one of joint interests and mobile ventures that shift at the speed of light. He does so by compellingly showing that one of the principal effects of artificial environments linked to populations (and not to a people) is that it becomes difficult, if not impossible, to locate a proper home for a population. Here Sloterdijk is mining another area of Foucauldian biopolitics on which few have touched, namely, the relation of biopolitics to chance and hence to security.[75] Indeed, Foucault speaks of aleatory events in the same breath as biopolitics: "a technology that brings together the mass effects characteristic of a population, which tries to control the series of random events that can occur in a living mass, a technology which tries to predict the probability of those events."[76] So, too, are Sloterdijk's immunitary designs anchored to individual (in)security completely inscribed here as well. Foucault thinks biopolitics as a "technology of security" that has to be installed around the random element inherent in a population of living beings so as to optimize a state of life. For Sloterdijk, the individual, and not the population, optimizes his state of life.

These areas of contact with Foucault demand a much fuller accounting than the one I am able to give here, but in the interest of moving the discussion forward, let me turn to where exactly Sloterdijk diverges with Foucault on biopolitics.[77] This occurs principally with respect to racism. We recall that for Foucault, racism was what made possible the reintroduction of death into biopolitics so that a separation could be introduced, a "break into the domain of life that is under power's control: the break between what must live and what must die."[78] The ultimate point of such an *aut aut* will be to allow power to treat populations as species. Where, then, do we find in Sloterdijk's analysis the mechanism that determines "what must live and what must die"? We have already seen a partial answer in Sloterdijk's earlier response, in which he makes the thanatopolitical implicit in the workings of the market through immunity. How so? Lurking behind Sloterdijk's reading of immunity, we find that for some to have immunity, to design their own immunity, others, the majority, must be stripped of theirs. He never says that for individuals to design theirs, others must lose theirs (as Esposito suggests in his reading of the *dispositif* of the person), but it is hinted at earlier, when Sloterdijk thinks a failing solidarity among one's own national collective that before would have offered forms of immunity to the larger community (immunity understood in its general sense as the welfare state and its forms of assistance). We see here again the role of neoliberalism in globalization, which cuts out from underneath community modes of collective protection. The process, emerging only in fragments in Sloterdijk, is one that Warren Montag has spoken of in terms of necroeconomics. For Montag, key in such a process is marking those whose could be allowed to die. As Montag writes,

> Thus alongside the figure of Homo Sacer, the one who may be killed with impunity, is another figure, one whose death is no doubt less spectacular than the first and is the object of no memorial or commemoration: he who with impunity may be allowed to die, slowly or quickly, in the name of the rationality and equilibrium of the market.[79]

Sloterdijk's analysis dovetails in many respects with Montag's, despite Montag's own interest in early-nineteenth-century economic thought—both locate in the market the means by which lives are abandoned. Sloterdijk, however, updates Montag's reading by linking it to failed communal regimes brought on by globalization, which becomes the ultimate step in which lives are thrown back literally onto their own devices.

The Thanatopolitics of Biotechnology (II): Globalized Gene Pools

Sloterdijk's more famous writings on biotechnology ought to be inscribed in this larger necroeconomic horizon. For Sloterdijk, the postmodern form of globalization enacted by capitalism cuts the ground out from below national immunity programs. Two results ensue. First, we move toward a "post-racist period, dominated by the evidence that the variations among individuals is in every case greater than those among ethnicities."[80] This is because globalization does not begin and end with the earth and that cosmic outside that turns us back on ourselves. It involves more significantly today a globalization of human gene pools. "This is characterized," Sloterdijk suggests, on one hand, "by an almost complete elimination of natural selection and, on the other hand, by a move toward the globalization of genes that will ultimately bring about a leveling of the historical differences between peoples and races [*Völkern und Rassen*]."[81] Sloterdijk reinterprets the move from people to population that figures so prominently in the Foucauldian narrative of biopolitics in the language of bioengineering. The troubling result is that as the difference between peoples falls as they move to populations, further differences will become so minimal as to remove the basis for racism itself. The question that Sloterdijk's analysis of globalization of genes raises is the following: if racism was the mechanism with which Foucault accounted for the appearance of the thanatopolitical at the end of the nineteenth century (later to become full-blown in Nazism), do we find ourselves now at the threshold of a different configuration of biopolitics, an anthropotechnical moment in which thanatopolitics has been done away with?[82] Sloterdijk seems to oscillate here. On one hand, as my earlier discussion of individuals without immunity protection suggests, he will incline the thanatopolitical toward a critique of the market, which will allow some lives to be abandoned. In this sense, he is clearly critical of neoliberalism and its inscription of life in death. On the other hand, in other texts in which biotechnology moves to the center, the market seems to disappear completely, vanishing in a Heideggerian haze of the mystery of Being. Thus being [*Sein*] prepares itself "for a continuous process of revelation and disclosure; for always new moments of suspicion and simultaneously, for hiding itself from that which now is manifest."[83] The reference, willed, it is clear, to revelation and discovery and then back to revelation, recalls those pages from Heidegger that have formed the subtext for my survey of the thanatopolitical in contemporary philosophy. Using our discussion of mystery and revelation as our angle for reading, we can say that Sloterdijk's perspective on biotechnology is also inscribed in technology and, with it, in the workings of proper and improper revelation and writing.

The emphasis on humanizing and bestializing media that opens "Rules for the Human Zoo" comes once again into play. For Sloterdijk, bioengineering functions like reading *(Lesen)*; in fact, he will read selection from the gene pool *(Auslesen)* as a form of reading. But bioengineering of genes is also implicitly a form of writing to the degree that it places us in a situation of mystery.[84] Biotechnology as writing does not reveal itself in one fell swoop but continues to hold man in place in a form of *Bestand* that I sketched in my discussion of Heidegger in chapter 1: in our being available to abandonment in biotechnology, we must wait for its effects to become apparent. This is the unspoken and unacknowledged economy of sacrifice in which some in the present are sacrificed for more efficient future immunitary designs. In other words, this individual immunization essentially does the work of death, pitting future bioengineering against contemporary terrorist attacks on immunity.

Where we might have expected Sloterdijk to argue that biotechnology is a form of improper writing that in fact withdraws man from Being, he will instead, in a reading that challenges *Parmenides* (or rather deploys the "Letter on Humanism" against it), deny a purely improper–proper reading to biotechnology. Bioengineering for Sloterdijk is radically different from other forms of technology (and hence other forms of writing) to the degree that the genetic engineering of superior individuals is now possible and is furthermore absolutely necessary—thus the still troubling pages in which Sloterdijk speaks of anthropotechnical governing:

> Royal anthropotechnology, in short, demands of the statesman that he understand how to bring together free but suggestible people in order to bring out the characteristics that are most advantageous to the whole, so that under his direction the human zoo can achieve the optimum homeostasis. This comes about when the two relative optima of human character or warlike courage and philosophical humanistic contemplation are woven together in the tapestry of the species.[85]

Sloterdijk will not simply note that technology standardizes mankind as a form of communication would, making us all the same: his argument about the coming period of postracism makes clear.[86] He will, however, also point out that mapped gene pools imply, as well, optimized gene pools. His unsettling reading of Plato shorn of its Nazi appropriations comes into view.[87] In these anthropotechnical techniques, Sloterdijk will see the ultimate attempt to deal with the aleatory element in life on the level of the individual. These individuals will become guardians of the human zoo so that in some future, the mass of humanity will no longer be different, except for those superindividuals whose superior gene pool permits them to act as the ultimate protectors of the majority of human animals. Again, we have another instance of life implicitly being strengthened through the death of others. Man waits for the proper mystery of his genetic future to be revealed, which places him in the position of requiring saving (even if Sloterdijk denies or limits this, given his deep desire to see bioengineering as heralding a new posthuman future). It is this figure of humanity who waits to be surpassed (and in so doing) who risks obsolescence—who is abandoned to the future.

We note, too, the implicit messianism of his analysis. We saw this in Agamben as well as Esposito, but differently from them, Sloterdijk's messianism is profoundly thought through a philosophical critique of technology; technology puts humanity at risk but will also save humanity by creating superior human beings who can manage the human animal zoopolitically—who can, through superior genetic pools, save the rest of us. What Sloterdijk cannot see is that technology in the form of bioengineering is also the product of the market and that the majority of humanity that he sees as having been exposed with little or no immunity protection risks being extended dramatically to all but those few individuals who have been engineered to rise above the animalized masses. Sloterdijk does not say much about these future biopolitical sovereigns, but their difference in species from the majority of humanity suggests again the role of the negative as well, because presumably, their own immunity designs will require, in not different fashion from the previous biopolitical sovereigns of colonialism, the distancing from the "community" of the human animals from which they are genetically immunized.[88]

To review, then, there are four moments of the thanatopolitical that characterize Sloterdijk's philosophical thought. The first appears in the final stages of globalization itself, when Western civilization, as he calls it, faced with a reconfigured and essentially nihilist notion of the outside as radically negative and other, is forced to view the earth from a certain immanent perspective. The earth itself becomes the subject of an immanent critique, without any transcendental coordinates for a homeland outside of the community to which the individual belongs. The second occurs with particular fury in European colonialism, when the Outside is related semantically to the other that is to be colonized. Colonialism in Sloterdijk's reading thus functions as a way of perfecting immunitary regimes, and so the thanatopolitical will be registered in the proliferation of ever more powerful immunitary *dispositifs* for the colonizer and the attempts on the part of the colonized to weaken them (thereby paradoxically strengthening them). The third we can see most clearly in the development of the *dispositif* of the household as a further instance of immunity—the design of immunitary regimes held privately and the weakening of immunity formerly provided by the larger community of the Nation-State. Here more explicitly than in the first is the thanatopolitical identified as the simultaneous expansion of individual immunity and the deimmunization of communities, in which one is premised on the other and with the death by exposure of others. The final moment occurs when Sloterdijk reads impure rage as the basis for a Marxist thanatopolitics, one premised on a future humanity of the proletariat and the implied homicide of those elements, namely, the lives of the bourgeoisie that do not participate in authentic forms of production.

By way of conclusion, let me backtrack slightly. I have awarded particular importance to technology in the emergence of thanatopolitics as the principal inflection of biopolitics in a number of important contemporary readings. I have done so since I see it uniting the three philosophers I have discussed throughout this work. But I note as well that it is technology as understood in the Heideggerian distinction between proper and improper writing and revelation and, in particular, the effects of technology, namely, its withdrawal of Being from man. For Agamben, we see this most clearly in his recent writings on the administration of life through increasingly advanced technology. Implicit for Agamben is a rush toward catastrophe that runs parallel to the increasing abstraction of intersubjective relations caused by communication technology as well as, implicitly, biotechnology. Underlying his anxiety is a fear of technology and its effects on a proper relation to others as well as the effects of "ordering," the folding in on those who attempt to master technology. For Esposito, even if the Heideggerian inflection of his reflections appears to be less pronounced than either Agamben or Sloterdijk, the anxiety of an improper revelation brought on by the invasive nature of technology becomes the pivot around which a politics of life can be envisioned, one that moves 180 degrees from the thanatopolitical through a more vitalist trajectory of philosophical thought that will include Spinoza, Nietzsche, and Deleuze. In Sloterdijk, we have the most fully developed branch of thanatopolitical reflection today. Moving across vast swaths of Western and non-Western history by adopting a sagittal gaze of the sphere, the thanatopolitical comes into view just as community begins to decline. Implicit in Sloterdijk's thanatopolitics is the idea that the individual is most threatened by technology, where technology has been expanded, thanks to the inflection Sloterdijk will give it as media, to include a host of *dispositifs* that run from the political party to the household to improper forms of rage to, finally and most importantly, the daily construction of indifference in media-driven democracies. Where there is technology, there is thanatopolitics and the need to strengthen individual life through the death of community. Sloterdijk registers this sheer explosion of immunity regimes and the concomitant exposure of millions to the threat of not being protected from capital. This does not lead him, however, to a full-fledged critique of neoliberalism and its tactics of depersonalization. Instead, he moves to a kind of blind grasping at technology itself—a new form of biotechnology by which humanity can begin to administer its own zoological features, when it can administer its own life through death.

4 PRACTICING *BÍOS*

Attention and Play as Technē

IS THE DRIFT TOWARD THANATOS the only possibility for contemporary forms of technologized existence? With increased technologization and its contamination with apparatuses, is *bíos* now pursuing a "decisive tack" in which "any residual hint of the anthropological is abandoned in the fact that techn-ology becomes properly speaking a thanato-logy?"[1] And with greater thanatology, must our responses be measured only in terms of managed births, as Sloterdijk polemically suggests, guarded over by Platonic human zookeepers?

In this chapter, I want to think another possibility for *technē* and *bíos* that will require repositioning *technē* as a practice able to configure different forms of life as forms of play. To set the scene, I want to return to that figure who both traced the thanatopolitics in biopolitics across a number of lectures from the 1970s but also attempted in the last years of his life to think through the aporia of *technē* and thanatos. I am speaking, of course, of Michel Foucault. My reasons for doing so will be obvious enough. More than any other thinker examined thus far, Foucault responds to the major questions that have characterized this study; indeed, in some sense, Foucault responds to himself across his later work—a thanatopolitical Foucault who is met by and, in my view, superseded by an affirmatively biopolitical Foucault. Bringing the two Foucaults together has the benefit of providing a launching pad for the heart of the chapter, namely, imagining an affirmative practice of attention and play in lieu of a thanatological *technē*—a different way of thinking *technē* not linked primarily to a defense of the self and its borders but rather as an opening toward the relational. I will conclude with a number of details about what a practice of *bíos* might look like by introducing at various moments those thinkers who have attempted to do just that. I end by suggesting that a practice of *bíos* might well move us toward what Nietzsche refers to as a "planetary movement."

The Pertinent Lives of Population

I have had occasion to discuss elsewhere some of the issues surrounding contemporary appropriations of Michel Foucault's term *biopolitics* as it appears in *History of Sexuality Volume 1* as well as his 1970s seminars, particularly *"Society Must Be Defended."*[2] Recently, however, another fold of Foucault's thinking of biopolitics and thanatopolitics has become clearer in the recently published seminar Foucault gave in 1978 titled *Security, Territory, Population.*[3] Here Foucault, just as in *History of Sexuality Volume 1*, defines biopower as "the set of mechanisms through which the basic biological features of the human species became the object of a political strategy or . . . how . . . modern Western societies took on board the fundamental biological fact that human beings are a species."[4] Differently from that earlier text, however, Foucault extends his analysis of biopower from the end of the eighteenth century through the nineteenth century, as human beings increasingly become an object of political strategies. He does so by leaning on Jean-Baptiste Lamarck's notion of milieu as a space in which populations are managed as species.[5] Foucault's first two lectures are of particular significance as he carefully maps how the biopower of populations is maintained and increased in a milieu, particularly in what he will call the passage from scarcity to scourge that occurs at the end of the eighteenth century. Out of his reading of milieu and population, the greater rationalization of governing populations comes into view, one linked to the science of statistics as well as advances in calculated management that make it possible for governments to manage scarcity. The primary intention of government, Foucault tells us, is to manage scarcity in such a way that it does not become a collective scourge. In the following passage concerning rationalization and scarcity, Foucault sets out the stakes of this decisive moment:

> But we will have an absolutely fundamental caesura between a level that is pertinent for the government's economic-political action, and this is the level of the population, and a different level, which will be that of the series, the multiplicity of individuals, who will not be pertinent, or rather who will only be pertinent to the extent that, properly managed, maintained, and encouraged, it will make possible what one wants to obtain at the level that

> is pertinent. . . . The population is pertinent as the objective, and individuals, the series of individuals, are no longer pertinent as the objective, but simply as the instrument, relay, or condition for obtaining something at the level of population.[6]

One immediately hears echoes of the larger biopolitical horizon that Foucault had discussed in the previous year's *"Society Must be Defended,"* in which life is both fostered and disallowed. Equally, though, one notes the descriptor *pertinent* and the deathly cast it takes on in the context of population: some lives will be more "pertinent" than others, based on whether they belong to a population or to a "multiplicity of individuals." In the "fundamental caesura" enacted by governments, the beginnings of an implicit thanatopolitical drift of life toward death appear to the extent that for the population to become the object of scarcity management, another series of "no longer pertinent objects," that is, individuals, will also be constructed as objects of (bio)power. Foucault's announcement of a caesura implies that the concerns of the formerly pertinent are heard less and, equally, that an economy is created in which multiplicities of individuals are used to manage populations—all premised on the opposition to individual difference. This distinction between multiplicity and population will later become critical when, in the midst of a crisis, one must manage not only scarcity, food principally, but more acutely security.[7] At this juncture, security comes into its own as the ultimate aim of government, when power moves away from simply managing populations to treating scarcity as an event in the larger horizon of security. Thus "the scarcity scourge disappears, but scarcity that causes the death of individuals not only does not disappear, it must not disappear."[8] Foucault assumes that as security comes to be a managed event, it is increasingly necessary to govern populations by never allowing the death of individuals to go unnoticed.[9] This introduction of security around the withdrawing figure of scarcity interests Foucault and should interest us as well, since Foucault is suggesting that for the population to be secured, doing away with scarcity is precisely what is *not* required. Instead, scarcity, understood as managing the death of multiplicity, is crucial for the administration of the population. For Foucault, we must not lose sight of the fact that managing populations is not to be thought apart from strategies of managing death. A first drift toward the thanatopolitical in Foucault's argument appears here, one that mirrors a number of prior moments in this study. The need for managing scarcity creates an opening for the thanatopolitical when scarcity, as an event to be managed, is made homologous to the death of individuals. Although Foucault does not detail the features of managing scarcity in linking security to population, he suggests that the death of individuals provides an access road for thinking an originary thanatopolitics folded into the figure of population. Furthermore, the individual exists in Foucault's analysis as both the mode by which the population is secured and the mode by which population is explicitly linked to thanatos.

Much more remains to be said about the relation between the death of the individuals and the securing of the population, in particular, how such a management strategy vis-à-vis populations continued to be elaborated across the last two hundred years. I have in mind the advent of the society of the spectacle and the birth of visual culture in Europe at the end of the nineteenth century as well as the appearance of communication networks after World War I and their respective roles in making certain that scarcity never disappears.[10] Taken together, they signal that managing events of scarcity through their spectacularization rehearses the fundamental caesura between population and individual that lies at the heart of the thanatopolitical in Foucault's thought.

Secure Circulation

Yet Foucault does not see the managed events of scarcity as the lone possibility for biopower and its object, the population, which is to say that the emergence of biopower and population together at the end of the eighteenth century is not a completely original context. In this regard, consider the return of the figure of the people in Foucault's analysis as a latent political possibility:

> Here too, in this sketch that begins to outline the notion of population, we see a division being made in which the people are generally speaking those who resist the regulation of the population, who try to elude the apparatus by which the population exists, is preserved, subsists, and subsists at an optimal level. The people/population is very important.[11]

Foucault will say little about how a multiplicity of individuals is transformed into a people.[12] More important will be how resistance by a people results from the productive caesura of populations and multiplicities in such a way that evading what he calls apparatuses that "preserve" populations becomes possible. Foucault has some specific ones in mind for obtaining "something that is considered to be pertinent in itself because situated at the level of population."[13] Many of these elements will involve "production, psychology, behavior, the ways of doing things of producers, buyers, consumers, importers and exporters, and the world market." "Security," he will go on, "therefore involves organizing, or anyway allowing the development of ever-wider circuits."[14] For Foucault, the security of the population does not reside only in the negative motor of individual death brought on and maintained by scarcity; rather, security is equally productive in the sense that in the development of circuits linked to capitalism—circuits of greater exchange between persons, primarily—security will come to be seen as profoundly connected to instances in which the "members" of a population are joined in circuits of exchange—these circuits substituting in some sense for the bonds (of language, of tradition) linking a people together. In other words, population and not people is to be thought as coterminus with the securing of exchange as a substitute for the ties among a people.[15]

In the conjunction of security with an apparatus that makes possible the development of circuits of exchange, we find a notion of *technē* returning exactly at a moment of nascent globalization. It occurs, in ways we should add, not so far removed from the readings that both Agamben and Sloterdijk will give of globalization. Aside from those affinities, let's simply note at this juncture the effects of this thinking of security and population together on the notion of freedom and, in particular, that of person because it is here that another fold in Foucault's understanding of the thanatopolitical will be found. Thus, in the same lecture, he will go on to note that "freedom is nothing else but the correlative of the deployment of apparatuses [*dispositifs*] of security," adding that freedom is "no longer the exemptions and privileges attached to a person, but the possibility of movement, change of place, and processes of circulation of both people and things."[16]

Two points need to be made straightaway. First, understood in this conjunction of *dispositifs* and capitalist circuits of greater exchange is a moment in which the form of person changes to reflect the loss of privilege and exemption brought on by the "development of ever-wider circuits." In a movement of increased circulation of goods, the form of person will change, and although Foucault does not return to the question of what the form of person might resemble after the *dispositifs* of security deexempt and deprivilege them, he intimates that an inverse proportion exists between securing populations and deprivileging persons.

And though in the intensification of the *dispositifs* of security, changes in the makeup of the previous person are implicitly gestured toward, Foucault leaves unexamined the nature of these changes. Yet, if we look more closely, an opening in Foucault's thought toward the thanatopolitical appears, one that has characterized all the principal protagonists of the following study. In the wider circuits of exchange that move hand in hand with the emergence of populations, the form of person is altered. Indeed, Foucault implicitly introduces the spectral form of the person in *Security, Territory, Population* and joins it to a key moment in economic and political liberalism. Foucault's analysis of population, security, and freedom, therefore, offers a powerful way of inscribing contemporary forms of technology within a horizon of biopolitics by affiliating forms of technology—communication, biotechnology, and bioengineering—with an intensification of the circulation of goods and persons. In fact, in most of the hymns to neoliberal genetics, the freedom to decide the qualities of a future human life through biotechnological processes is often located within this "option of circulation." The result is to link such changes in the understanding of personhood with the widening network of circuits of exchange of genetic material. Securing populations from future disease and harm continues to remain the ultimate context for such considerations.[17] This is the horizon in which Aihwa Ong speaks of Singapore's recent and successful attempts at creating laboratories dealing in genetic material and a new kind of "biosociality" that demands the domestic population "to turn against its own deeply held beliefs . . . and yield up genetic tissues for transreligious commingling."[18] The conclusion? Contemporary neoliberalism, under cover of greater circulation and freedom, brings in its wake a machinery whose job it is to secure the same circulation through a resort to a sort of sovereignty inscribed in biopower and political authoritarianism. Here, too, the threat of depersonalization as populations increasingly are seen as mere reserves of genetic tissue is real. What communication technologies and genetic engineering share, therefore, in a neoliberal, thanatopolitical regime of circulation and exchange is their capacity to ease circulation globally, while at the same time deploying an apparatus meant to make disposable these former persons as a means for securing the very same populations of which they are a part.[19]

In this coterminus move toward population, security, and territory, not only does the category of person change such that exemptions and privileges are torn from her, but just as decisively, other collective political figures, in particular, the community, come under siege. Population in such a Foucauldian perspective comes increasingly to appear as the anamorphosis of community in ways that mimic the other figures I have examined here. Admittedly, Foucault never explicitly contrasts populations with community in *Security, Territory, Population.* Yet, by summing security and population together into territory, Foucault shows how forms of collective life give way to something else under the increasing domination of biopower and its *dispositifs.* As national or local communities recede and mobile populations gain ground, the former political space of community, with its awarding of protections and privileges, is weakened. The result is not just a "detachment of entitlements from political membership and national territory, as certain rights and benefits are distributed to marketable talents and denied to those who are judged to lack such capacity or potential," but something even more dramatic: the descent of community as a political and spatial category into mere territory.[20] Massimo Cacciari recently mused on urban space and territory in this regard, positing that "we no longer live in cities but rather territories (territory from *terreo,* to have fear, to experience terror)."[21] As the reader knows by now, the shift to the terrifying spaces of noncommunity is at the heart of contemporary reflections on the thanatopolitical: the space of terror in territory occurs not only thanks to intensified security apparatuses of the kind Foucault noted in terms of scarcity but also because technology, according to the protagonists of this study, accelerates the move to territories of nonpersons or semipersons away from community. The resulting space will paradoxically be one in which a secured and free (free precisely because secure) population inhabits a territory—populations that consist in turn of varying degrees of personhood. In and out of this depersonalized, territorial space, events of scarcity will be managed and, with them, decisions taken with regard to those individuals who are deemed to be less than pertinent. That is the darker possibility Foucault suggests in *Security, Territory, Population,* in which security is practiced on territories inhabited by populations. In the increasing expanse of circuits of exchange, territories grow out of (or are grafted onto) former national and local communities. These territories comprise not only "full-fledged" members but also include within their ranks those nonpertinent individuals whose lives and deaths are increasingly managed and so made scarce.

Such a reading of technology, population, and biopolitics challenges those who continue to ignore the role of security in relation to neoliberalism and globalization.[22] Frequently, my impression is that a number of critics avoid a real grappling with Foucault's insight here into the relation of neoliberalism, freedom, and an implicit thanatopolitics. To the degree that technology operates in a fashion similar to those earlier instances of an intensification of biopower linked to greater circulation of goods, we continue to live in and under the sign of increasingly secured populations; in fact, given the intensification of ever-widening circuits of exchange, nothing seems to suggest that the biopower associated with populations has diminished. Indeed, biopower continues to grow exponentially. In *The Birth of Biopolitics,* Foucault's follow-up to *Security, Territory, Population,* he uncovers the hidden assumption of the market's rationality and equilibrium in the notion of population itself, as that which contains within it a mechanism of scarcity by which individuals may not only be allowed to die but to be pushed over into death.[23] Foucault's superimposition of the market on population allows us to see how neoliberalism, in its "defense" of freedom, depends on an intensification of security through widening circuits of exchange.[24] Such an overlapping forms one of the principal ways in which Foucault's thinking of population and security may be inscribed in a thanatopolitical horizon. Such a reading of *Security, Territory, Population* has the advantage of making visible the fault lines of the intersection between the emergence of biopower and security with forms of *technē* that are developed for managing pertinence and so with some of the principal figures of this study thus far. One might even go so far as to suggest that a significant affinity can be found between Foucault's division between pertinent and impertinent and between proper and improper writing. At a minimum, it appears that Foucault draws our attention to the intensification of modes of securing populations as different forms of *technē* grow more pronounced.

Biopolitical Ethics

Is securing populations the only possibility for biopolitics in a technologized milieu, its increasing inscription as only biopower, with only a toxic mix of *dispositifs* and media to look forward to? As I pointed out in the introduction and suggested in those pages dedicated to Agamben's and Esposito's positing of an impersonal potentiality, I believe another possibility for biopolitics exists today, but it must be linked to another moment in Foucault's thought; not to the Foucault of *"Society Must Be*

Defended" or *History of Sexuality Volume 1* or *Security, Territory, Population,* but to a later Foucault, a "wounded Foucault," as Ida Dominijanni describes him, or more popularly, an ethical Foucault.

Unfortunately, many continue to use the ethical descriptor when discussing the readings of later Foucault to limit the organic links between a political Foucault of the clinic or the prison with a later Foucault principally interested in an ancient care of the self that would be fundamentally ethical in nature and hence removed from his earlier biopolitical perspective. Such a point of view has made it difficult to bring together biopower and these later interventions of Foucault, with biopolitics implicitly discounted when evaluating Foucault's later work. I prefer to see a key relation between Foucault's earlier reflections on biopolitics and these final considerations of his on the "care of the self"—to see Foucault's ethical perspective as a response to an earlier diagnosis of biopower. Thus, throughout the following pages, I will be suggesting that Foucault in these later works seeks to disclose a response to modernity's increasing biopower by introducing a form of technology, a *bíos technē,* as a possible response to his earlier research on biopower and thanatos. I will be reading Foucault's reflections on techniques for life as a response to the zoologification of life that has emerged in all the readings of the thanatopolitical to this point as originating out of a disjunction between proper and improper forms of life, between *bíos* and *zoē.*

In other words, I want to sunder the distinction between an ethical and a political Foucault in an always already biopolitical Foucault. To the degree that Foucault insists on a *technē* for *bíos* in *Hermeneutics of the Subject,* I attempt to imagine another form of *technē* for *bíos* that would be thought together with two other possibilities for *bíos*: attention and play. Both, in my view, might offer us a response to *technē*'s inscription in thanatos, a relation that continually moves through incorporation and expulsion. I will be drawing on a number of Sigmund Freud's essays in these sections, Gilles Deleuze's notion of haeccetic space, Guattari's views on creation, D. W. Winnicott's understanding of play in a therapeutic setting, and finally, Walter Benjamin's philosophical lexicon of playing and toys to sketch a practice of *bíos* that would avoid the problems of a care for the self that seemingly devolve into mastery in Foucault's reading. My project is to see in attention and play possibilities for weakening the borders of the self, which are continually reinforced by a *technē* that no longer has any relation to life. I hope to find in attention, therefore, a *technē* of *bíos* that avoids any complicity in proper and improper forms of life—that resists the division between *bíos* and *zoē* that a Heideggerian reading of *technē* seems inevitably to call forth.

To begin, what role does Foucault award biopower in the 1981–82 lectures collected in *Hermeneutics of the Subject*? The answer, which explains the initial critical reaction to the essays as part and parcel of a more ethical Foucault, is not very much. As far as I can make out, Foucault never once mentions biopower or biopolitics throughout the entire year; rather, his goal as set out in the first hour from January 6, 1982, is to write a history of a care for the self as a cultural phenomenon:

> What I would like to show you, what I would like to speak about this year, is this history that made this general cultural phenomenon (this exhortation, this general acceptance of the principle that one should take care of oneself) both a general cultural phenomenon peculiar to Hellenistic and Roman society . . . and at the same time an event in thought.[25]

Foucault will go on to associate a care of the self with a "critical ontology of ourselves" that will ultimately be joined to modern subjectivity:[26]

> It seems to me that the stake, the challenge for any history of thought, is precisely that of grasping when a cultural phenomenon of a determinate scale actually constitutes within the history of thought a decisive moment that is still significant for our modern mode of being subjects.[27]

Although Foucault does not detail further this "modern mode of being subjects," such a mode haunts the rest of the lectures, reaching a kind of denoument in the closing pages, when Foucault will associate it with a moment in thought in which *technē* moves out of the horizon of *bíos*, leaving *bíos* to be captured by the *technē* of the self. Thus not only is Foucault's interest in care of the self as broad as the history of thought itself but such a care cannot be separated from an ontology of the present linked to a seminal event in thought, when *technē* no longer enjoys a primary relation to forms of life.

In *Hermeneutics of the Subject,* Foucault spectacularly recapitulates a history of care of the self across Greek and Roman antiquity, from Platonism to Stoicism, providing numerous examples of therapies offered as the means by which individuals care for their selves. I want to focus initially on the conditions under which a care for the self develops because there we will find another division of proper and improper life. After noting how "in Greek, Hellenistic, and Roman culture, care of the self always took shape within quite distinct practices, institutions, and groups which were often closed to each other, and which usually involved exclusion from all others," Foucault observes that care of the self always took shape "within definite and distinct networks or groups, with combinations of the cultic, the therapeutic . . . and knowledge, theory, but [involving] relationships that vary according to the different groups, milieus, and cases."

We note immediately that care is not available to individuals simply on the basis of being human. On this score, Foucault states,

> If you like you cannot take care of the self in the realm and form of the universal. The care of the self cannot appear, and above all, cannot be practiced simply by virtue of being human as such, just by belonging to the human community, although this membership is very important. It can only be practiced within the group, and within the group in its distinctive character.[28]

I will return to this point later, when I turn to the notions of play and attention as well as the assumption such a view makes about what Leela Gandhi in a not-so-different context has called an "immature politics." Instead, let's note that Foucault does not simply register the affinities between a "technology of the self" in antiquity and Christianity, with their attending problems of revelation, faith, and grace, but also the tension that inheres in such a *technē* between being human as the essential condition in practicing a care of the self and the necessity of belonging to some larger group. A glance at the etymologies of *belonging* is helpful. According to the *Oxford English Dictionary, belonging* first meant "corresponding in length," which, over time, took on the additional meaning of "going along with, accompanying as property or attribute." The two etymologies come together to signify its final meaning of that which is "appropriate."[29] All these meanings circulate in the preceding passage. On Foucault's read, the care for the self depends on that form of life that had as one of its properties the quality of belonging. More noteworthy in my view is that Foucault never sees the mere fact of living in antiquity as a sufficient condition for a care for the self, leading him in the closing sections of his lecture of January 20, 1982, to discuss salvation and the problem of "what is it to be in good health, to escape from illnesses, both to be led to death and in a way to be saved from death?" Foucault finds in the tension between being led to death and being saved from it an implicit dialectic between a form of life in which care is possible to the degree one belongs to a group (or network, if we want to pick up the technological flavor of his argument) and a life linked to a mortality that is caused by the mere fact of being alive.

Foucault sees this movement between the universal and the particular, a nonbelonging of the merely human and a belonging appropriate to a group or collective, at the heart of the question of salvation. Certainly it is one a number of readers of the later Foucault have stressed. We, too, can see that Foucault in these lectures refuses a simple choice between them, which is to say, he does not opt for a quality of belonging, a proper life as opposed to an improper one, as Agamben does, whereby the improper one would be lined up with potential. Though the remainder of the lectures range widely across the various manifestations of askesis in Greek and Roman antiquity and the construction of a "complete relation of oneself to oneself," the tension between the universal and the particular, between belonging and being merely human, is maintained.

What does Foucault make of the relation between proper life and care of the self? Foucault moves in two directions simultaneously: on one hand, toward those *dispositifs* crucial for a care for the self and, on the other hand, toward those *technē* of *bíos* not completely homologous to such apparatuses of the self. Speaking across a number of important pages about the significance of the term *meditatio* in Latin (*meletan* in Greek) as associated with belonging to a particular group, he writes,

> First *meletan* is to perform an exercise of appropriation, the appropriation of a thought. . . . The *meditatio* involves, rather, appropriating [a thought] and being so profoundly convinced of it that we both believe it to be true and can also repeat it constantly and immediately whenever the need or opportunity to do so requires.[30]

Foucault describes this assistance as a *prokheiron* in Greek or in Latin as *ad manum* (ready to hand), an apparatus of truth that will serve "as an exercise for the day he suffers a misfortune, so he will have *prokheiron* (ad manum: ready to hand), the apparatus of truth which will allow him to struggle against this or that misfortune, when it arrives."[31] Foucault thinks modes of appropriating a thought through the notion of *dispositif* but does not link them to strategic relations of force primarily; rather, he joins them to a proper mode of care for the self.

Such a "ready at hand" echoes Heidegger's notion of improper writing and of not having at hand. The reader will recall, in particular, those pages in which Heidegger, in *Parmenides,* argues that a different kind of writing, an improper one linked to the typewriter, led to all sorts of disastrous results, in particular, to a thanatopolitics associated with the birth of the Communist technical man. Foucault, too, focuses on the notion of the hand in antiquity for a practice of the care for the self that is a part of another apparatus of truth. Foucault relates this having ready at hand to a practice of "listening, reading, and writing" that will allow the subject to say the truth about oneself, what he names in *Fearless Speech* as *parrhesia.* The practice of a care for the self cannot be thought in Foucault's view apart from the apparatus of the letter that is ready at hand, which can be reread and thus incorporated or memorized for future moments of misfortune. We can sum up Foucault's view on care of the self this way: the sense of belonging to a group as what gives proper form to life results from an apparatus of the letter that is ready at hand only to the degree that it originates in and from a collective form of being across time.

One further point: having these letters or writings as a *dispositif* ready at hand produces a kind of subjectivity for the philosopher, one that is intimately related to freedom. These are not monks of the Christian sort who are asked to follow rules but those in antiquity who put into practice the art of living as care of the self. In fact, Foucault will make a distinction between a Christian framework of a rule of life as opposed to antiquity's *technē tou biou* (an art of living). He writes, "Making one's life the object of a *tekhnē,* making one's life a work—a beautiful and good work (as everything produced by a good and reasonable *tekhnē* should be)—necessarily entails the freedom and choice of the person employing this *tekhnē.*"[32] Foucault will see a certain form of *technē* as being indispensable for the art of living. Yet a distinction between rules for life, on one hand, and art and living, on the other, appears here, and it is the *technē* that functions as the operator of difference between them.

A *Technē* of *Bíos*

What is the nature of the *technē* that is capable of making life its object? Foucault will devote many pages to this question in *Hermeneutics of the Subject,* but he comes closest to joining forms of life to *technē* in the lecture from March 17, 1982. Reading Seneca on abstinence, Foucault spells out the relation of forms of life to *technē*:

> In other words, what Seneca is aiming for in this kind of exercise is not the great conversion to the general life of abstinence, which was the rule

> for some Cynics and will of course be the rule in Christian monasticism. Rather than converting oneself to abstinence, what is involved is the integration of abstinence as a sort of recurrent, regular exercise to which one returns from time to time and which enables a forma (a form) to be given to life, that is to say, which enables the individual to have the appropriate attitude [towards] himself and the events of his life; sufficiently detached to be able to bear misfortune when it arises; but already sufficiently detached to be able to treat the wealth and goods around us with the necessary indifference and with correct and wise nonchalance.[33]

Foucault goes on to note how these exercises of abstinence have the express purpose of "forming a style of life" and are not "exercises of abstinence for regulating one's life in accordance with precise interdictions and prohibitions." The notion of style merits our attention, but for now, it is enough to note that Foucault relates abstinence and renunciation to giving form to life, as if the not fulfilling of desire effectively gave form to life. Indeed, something about practicing abstinence as part of a *technē* of *bíos* permits the individual to develop the proper attitude toward herself; a homology arises between the proper perspective on oneself and the organic link of giving form to one's own life. Indeed, we can see this in Foucault's later gloss of Seneca in the movement between the proper, the form of life given, and the mode of being subject, in which a sense of detachment from the events of one's own life, in particular, misfortune, is constructed. Abstinence as *technē* affirms a living to the degree one develops a proper relation to one's own life. For that to happen, it would seem that a distance from desire makes life ready at hand and hence proper to the individual.

A point requires clarification: throughout these pages, Foucault does not draw a clear-cut distinction between care of the self and *technē* of *bíos*, which is to say that though the latter is associated with a care of the self, it is not completely captured by the self. The self emerges, instead, as one among a number of possible forms of living or forms of life. In other words, Foucault does not make forms of life conditional on a mere care for the self. Some of the reason will become clear shortly, but for now, Foucault does not characterize the features of the care of the self by a mastery of the self by the self;[34] rather, he chooses to highlight freedom.[35] Of great interest, therefore, is that over time, the distinction between the two, abstinence and *meditatio*, as *technē* will decrease in importance, and instead, the test of the self will grow in Foucault's account. In subsequent sections of the lectures dedicated to the Hellenistic period, a distinction grows between *technē tou biou* and *care for the self* such that the care for the self that results from abstinence and other practices will be linked to the *praemeditatio malorum,* whereas a later care for the self will be thought primarily through the test.

This test, or what Foucault will call on occasion "self-questioning," creates difficulties for a *technē* of *bíos*. The *technē* concerns developing "a general attitude towards reality" such that the whole of life "must become a test."[36] Foucault writes,

> The care of the self is not something with which one must begin if one wishes to define properly a good technique of life. It seems to me that henceforth the care of the self not only completely penetrates, commands, and supports the art of living—but the *tekhnē tou biou* (the technique of life) falls entirely within the new autonomized framework of the care of the self.[37]

"What," he continues, "is the meaning and objective of life with its formative and discriminating value, of life in its entirely, seen as a test?"[38] The answer?

> It is precisely to form a self. One must live one's life in such a way that one cares for the self at every moment and that at the enigmatic end of life . . . what one finds, what anyway must be obtained through the *tekhnē* one installs in one's life, is precisely a certain relation to the self, which is the crown, realization, and reward of a life lived as a test.[39]

Foucault links *bíos* to existence as the object of plural *technē* and obliquely suggests that the notion of *bíos* ultimately changes across antiquity to become less a *technē* and more "the form of a test of the self."[40]

Many questions come to mind, but most important may well be whether there is a mirroring of the difference between *technē* of *bíos* and *technē* of the test in the difference between *technē* and *dispositif* (apparatus)? How might such a difference be productive for an affirmative biopolitics? Before turning to these questions, let's first mark the moments that together spell the development of *bíos* in Foucault's account. The first consists of the prior moment in which *bíos* results from making living the object of *technē* and whose practices include frank speech, *meditatio,* and abstinence. The second moment, which Foucault will contextualize on very nearly the last page of the year's lectures, occurs when *bíos* is no longer merely the object of *technē* but is now invaded by a care for the self that cannot be thought apart from mastery. Foucault describes the decisive moment of *bíos* moving out of the realm of a more ecumenical *technē* in the following way:

> Now if we accept . . . the idea that if we want to understand the form of objectivity peculiar to Western thought since the Greeks we should maybe take into consideration that a certain moment, in certain circumstances typical of classical Greek thought, the world became the correlate of a *tekhnē*—I mean that at a certain moment it ceased being thought and became known, measured, and mastered thanks to a number of instruments and objectives which characterized the *tekhnē*, or different techniques—well, if the form of objectivity peculiar to Western thought was therefore constituted when, at the dusk of thought, the world was considered and manipulated by *tekhnē*, then I can think we can say this: that the form of subjectivity peculiar to Western thought . . . was constituted by a movement that was the reverse of this. It was constituted when the *bios* ceased what it had been for so long in Greek thought, namely the correlate of a *tekhnē*; when the *bios* (life) ceased being the correlate of a *tekhnē* to become instead the form of a test of the self.[41]

A change in the fortunes of *bíos* occurs when the test becomes the dominant mode by which care for the self is practiced in a framework that Foucault will call "autonomized." Perhaps *technē* is not the proper term here either because what results from it is a mastery over the self. My impression is that for Foucault, such a dominating role for the test is something to bemoan because the test, so integral to a later and limited care for the self, shifts the ground out from under *bíos* such that *bíos* now merely stands in as homologous to the self. At the same time, another change takes place in the relation between *technē* and the world. In a kind of mobile overlapping, *technē* moves outside the domain of life to the world such that *technē* as a subject of *bíos* now becomes the subject of the world. A possible conclusion is that introducing *technē* outside any link to forms of life that reside outside the self leads to mastery over the world and, at the same time, to the distancing of *technē* from forms of life. Furthermore, it appears that more than one form of life prior to *technē*'s independence was available when *technē* was associated with constructing different forms of life not linked solely to the self. What cannot be doubted, however, is that at the moment when care for the self as test is born, *bíos*'s relation with *technē* changes. The assumption is that the test is either categorically different from *technē* or is *technē*'s degraded form.[42]

Foucault's disavowal of *technē* for *bíos* has a number of consequences. First, it encourages us to disagree with those, such as Agamben, who continually collapse all *technē* into *dispositifs*, whose sole purpose is to capture subjectivity so as to envelop it (and by extension, humanity) in a ruined and impolitical process of desubjectivization. On Foucault's read, however, *technē* is not simply a *dispositif*, given that *technē* wasn't historically always interested in capturing *bíos* as merely the self, as a mode to master care; rather, *technē* formerly was the impetus for the construction of forms of life.[43] It may sound strange to our ears, but the withdrawal of *technē* from life opens a space for greater mastery of the self and, so doing, closes off other possibilities for *bíos* as well as the self. Second, the distinction between a *technē* of *bíos* and a testing of the self is not mirrored in the biopolitical distinction between *bíos* and *zoē* because those who care for themselves through testing do not in anyway resemble a life unqualified by any attribute, political or otherwise. When seen against the background of some of Foucault's later comments offered in interviews as well as the general overview he gives of the lectures, the passage suggests that mere care for the self as a test both accounts in some way for our "modern mode of being subjects" and yet fails to live up to the demands of that mode. Perhaps in the unavailability of *technē* for life outside the self, we have something like a hidden genealogy of biopower. What itineraries can we imagine that allow us to elaborate a *technē* up to the demands for rejoining *technē* with *bíos*? What *technē* are available that would allow us to break the hold that the self as test continues to have over *bíos*? In different words, can we imagine *technē* today as a practice of *bíos* that might lead to forms of life that are not specifically limited to the self and mastery over it? What would they look like, and why would we even hold out hope that they might be able in some way to measure up to the stunning thanatopolitics of *technē* over life itself that has been the theme of this study to this point?

Let's recall just how deeply involved Foucault was near the end of his life in thinking the relation of self to *bíos*. In *Fearless Speech* as well as a series of interviews at the time, Foucault looked forward to breaking out of an equivalency between life and self. Foucault confirms such a reading in an interview given at the same time as these lectures at the Collége de France:

> What I want to show is that the general Greek problem was not the *tekhnē* of the self, it was the *tekhnē* of life, the *tekhnē tou biou*, how to live. It's quite

> clear from Socrates to Seneca or Pliny, for instance, that they didn't worry about the afterlife, what happened after death, or whether God exists or not. That was not really a great problem for them: the problem was: Which *tekhnē* do I have to use in order to live as well as I ought to live? And I think that one of the main evolutions in ancient culture has been that this *tekhnē tou biou* became more and more a *tekhnē* of the self. A Greek citizen of the fifth or fourth century would have felt that his *tekhnē* for life was to take care of the city, of his companions. . . . With Plato's Alcibiades, it's very clear: you have to take care of yourself because you have to rule the city. But taking care of yourself for its own sake starts with the Epicureans—it becomes something very general with Seneca, Pliny, and so on: everybody has to take care of himself. Greek ethics is centered on a problem of personal choice, of the aesthetics of existence.[44]

Such a move to *bíos*'s relation to art runs parallel with an earlier moment in *Hermeneutics of the Subject,* in which Foucault discusses the difference between rules of life and the art of living, the *technē tou biou*. There he writes, "Making one's life the object of a *tekhnē*, making one's life a work—a beautiful and good work—(as everything produced by a good and reasonable *tekhnē* should be) necessarily entails the freedom and choice of the person employing this *tekhnē*."[45] When juxtaposed to the interview, we see that Foucault literalizes a *technē* of *bíos* as an art of the living. Thus "a beautiful work is one that conforms to the idea of a certain *forma* (a certain style, a certain form of life)."[46] We note, too, that the turn toward the Greeks for a *technē* of life also implicitly meant for Foucault a genealogy of biopower. When asked in the same interview whether it wasn't logical "that you should be writing a genealogy of bio-power," Foucault responds ruefully, "I have no time for that now, but it could be done. In fact, I have to do it."[47]

What emerges in such a reading of Foucault is therefore, on one hand, a possible genealogy of biopower linked to self's enmeshment in mastery and, on the other, aesthetics as offering *bíos* a mode for responding to biopower. In this regard, consider Foucault's enigmatic suggestion, offered in one of his final interviews, of positing *bíos* as a material for aesthetics: "The idea of *bios* as a material for an aesthetic piece of art is something that fascinates me."[48] Indeed, he will speak of "*bios* as a material for an aesthetic piece of art." Certainly one of the mainstays of modern Western thought and literature is the idea that one's life can be turned into a work of art; fin-de-siècle, art for art's sake later gave birth to a vital life for life's sake, which characterizes so many of the early-twentieth-century avant-garde from Hugo Ball to F. T. Marinetti to Andrè Breton. Foucault surely recognized these modern precursors in his thought. Yet how would a conjunction of aesthetics and *bíos* function as an alternative to a *technē* of the self as test? Equally, what kind of material is *bíos* such that it can be transformed into a vitally aesthetic piece of art (or an aesthetically vital piece of art)? Apparently, aesthetics' more direct manipulation of *bíos* (and not the self) differs from the mastery of the self that characterizes heroic modernism and the avant-garde in particular (because presumably, there the self remains primarily the object of aesthetic *technē*).[49] A *technē* of *bíos* thought through aesthetics for Foucault might be one better equipped to disconnect life from the intensification of power relations that takes place increasingly under biopower.[50] To gloss the end of Foucault's *Fearless Speech*, an aesthetics of *bíos* would involve taking up the role of "a technician, of a craftsman, of artist" toward *bíos* itself and not the self.[51]

The Self and Biopower

Some of this technician of *bíos* appears in two seminal readings of the self and biopolitics that appeared after Foucault's death. Both Donna Haraway and Judith Butler emphasize the self's complicity with the intensification of power relations. Haraway forcefully places such an entanglement of the self with biopower front and center in war discourse when she asks, "When is a self enough of a self that its boundaries become central to entire institutionalized discourses in medicine, war and business. Immunity and invulnerability are intersecting concepts." Her conclusion? "Life is a window of vulnerability."[52] So, too, Judith Butler. In her reading of recognition protocols for the self that move through the Other, she notes as well that the self and, with it, self-preservation cannot be the highest goal and that the "narcissistic point of view" is not the most urgent need.[53] She winds up where Haraway does: with the explicit recognition that any process that reinforces a certain form of the self will continue to posit an ontological difference between self and Other, or in her terms, judge and judged.[54]

Yet neither Haraway nor Butler ultimately faces the question that Foucault's discussion of the self raises: are all modern attempts at a care for the self complicit in *technē*'s mastery of the world (and implicitly mastery of *bíos* as well)? Butler and Haraway come right up to the question of the self and preservation but fail to highlight sufficiently the role that *technē* plays there. In Butler's case, this means that her emphasis on the ethical Foucault read through Levinas fails to inscribe the later Foucault in questions of biopower and hence fails to ask a fundamental question: are there forms of life that do not primarily move through the self that are better equipped to resist such a mastery of the world? Haraway's larger point about the cyborg and a space of hybridity deals better with the self's enmeshment in discourses of war as she attempts to slacken the relation between power and cyborg. But here, too, the question of *technē*'s relation to *bíos* remains open.

Why the hesitancy? Surely some of the reason concerns the failure to think the self and its *technē* of the test as in any way different from previous *technē* of *bíos*. What is at issue is how the mastery of the self and the mastery of the world have as their result an intensification of the borders of the self, premised on a limitation of *technē* of *bíos*. How, we might well ask, does an intensification of the self's borders work? For a possible answer, I want to introduce Freud's reading of the self and what he calls the "self-preservation" drive or instinct. To do so offers us a preliminary step to where I ultimately am heading, namely, imagining possible practices of *bíos* in lieu of self *technē*. That we should introduce Freud in a context of *technē* and biopower will surprise no one, given the role Freud awards negation in terms of self-preservation. Let's consider a number of passages from Freud, all coming from two seminal essays.[55] The first is taken from "Drives and Their Fates." Here Freud, in an attempt to set out the difference between "ego or self-preservation drives" and "sexual drives," sketches the process whereby the original ego comes into being:

> The ego does not need the outside world, but, as a result of experiences undergone by the self-preservation drives, it does acquire objects from it. . . . So under the rule of the pleasure principle another development now takes place. The ego takes the objects it encounters, in so far as they are sources of pleasure, into itself, it introjects them . . . while, on the other hand, expelling whatever within itself causes unpleasure. . . . The original reality-ego, which distinguished an inside from an outside by means of a sound objective criterion, thus turns into a purified pleasure ego, which puts the factor of pleasure above all else. The outside world is divided up into a pleasurable part, which it incorporates into itself, and the rest, which is alien to it. It also separates off a part of its own self, which it projects into the outside world and perceives as hostile.[56]

The "sound objective criterion" for Freud is protection from stimuli arriving from the outside and the ego's instinctual move away from them.[57] Implicit in his analysis is that attachment to pleasure already introjected cannot be separated from the stimuli arriving from outside. This leads Freud to the following stunning conclusion:

> If the object is a source of unpleasurable sensations, there is an impulse to increase the distance between it and the ego, repeating the original attempt at flight from the stimuli of the outside world. We feel "repulsion" at such an object and hate it; this hate can then escalate into an aggressive inclination towards the object, an intent to destroy it.[58]

The conclusion? "As an object relation, hate is older than love, its source being the narcissistic ego's primal rejection of the stimuli of the outside world. . . . It [hate] remains forever closely related to the self-preservation drives." In Freudian terms, the "problem with the self" cannot be thought except as part of a systematic process of incorporation and expulsion; it is for this reason that love for Freud at its preliminary stages is scarcely distinguishable from hate. In each instance, the self-preservation drives dominate over what Freud will call the sexual drives.

Freud examines the division of inside and outside on which the self's preservation is based in another seminal essay, "Negation." Here Freud superimposes judgment over the original reality ego's relation to the outside and spells out the relation between stimuli and preservation more fully:

> The function of judgment is concerned in the main with two sorts of decisions. It affirms or disaffirms the possession by a thing of a particular attribute; and it asserts or disputes that a presentation has an existence in reality. The attribute to be decided about may originally have been good or bad, useful or harmful. Expressing the language of the oldest—the oral—instinctual impulses, the judgment is: "I should like to eat this," or "I should like to spit this out"; and, put more generally: "I should like to take this into myself and to keep that out." That is to say: "It shall be inside me" or "it shall be outside me." As I have shown elsewhere, the original pleasure-ego wants to introject everything into itself that is good and to eject itself everything that is bad. What is bad, what is alien to the ego and what is external are, to begin with, identical.[59]

If we were to follow out the process of introjection and expulsion, the self's attachment to objects seemingly begins with an originary division in which what is incorporated cannot be separated from what has been expelled: to be on the outside is already to have been expelled and hence to be alien to the ego. In this sense, the self's relation to objects as stimuli cannot be thought apart from the defense of the inside, that is, introjection and a moving outside (projection).[60] Interestingly, Freud will go on to relate the process of introjection and expulsion that was central to the self-preservation drive of "Drives and Their Fates" to the "instinct for destruction." "The polarity of judgment," he notes, "appears to correspond to the opposition of the two groups of instincts [drives] which we have supposed to exist. Affirmation—as a substitute for uniting—belongs to Eros; negation—the successor to expulsion—belongs to the instinct [drive] of destruction."[61] As many have noted, some similarity, or even perhaps an anxiety of influence, is visible in such a perspective, which evokes Hobbes and the state of nature, given the emphasis Freud places on the "ego's struggle to preserve and assert itself."[62] I prefer to focus alternately on the polarity of judgment that is inscribed in the self-preservation drive and, with it, the drive for destruction because here we find an opening to the negative of judgment, or better, the negative of a negation.[63] How do we want to speak of the negative of judgment? One way, though by no means the only one, is as a form of standing in place, a state that does not immediately move toward expulsion or incorporation, something in line with the final pages of *Fearless Speech,* in which Foucault speaks of avoiding taking up the "role towards oneself as that of a judge pronouncing a value."[64]

Something else deserves mention. Implicit in the reading of *technē* and *dispositif* across the previous chapters was something like a negative inscription of *technē* that is always already located in the drive of destruction. We sense it in Heidegger's technologized man, who is prepared for (and prepares) destruction, and we see it in Sloterdijk's raging political parties. One of the merits of Freud's reading of the drive for destruction will be located in having unearthed a vector of thanatos in *technē*'s reinforcement of the self by providing the self with extensions as a way of moving outside and of incorporating.[65] The result is one that the Invisible Committee recently analyzed:

> "I am what I am." My body belongs to me. I am me, you are you and *something's wrong.* Mass personalization. Individualization of all conditions—life, work and misery. Diffuse schizophrenia. Rampant depression. Atomization into fine paranoiac particles. Hysterization of contact. The more I

> want to be me, the more I feel an emptiness. The more I express myself, the more I am drained. The more I run after myself, the more tired I get. We treat our Self like a boring box office. We've become our own representatives in a strange commerce, guarantors of a personalization that feels, in the end, a lot more like an amputation. . . . Meanwhile, I *manage*. The quest for a self, my blog, my apartment, the latest fashionable crap, relationship dramas, who's fucking who . . . whatever prosthesis it takes to hold on to an "I."[66]

In this interaction between prosthesis and *technē*, we continue to hold on to the self—a self that is strangely empty and one that continues to be the object of loss or amputation. Such a diagnosis would have prosthesis and judgment dovetail; judgments provide a way of holding on to a self and thereby reinforce a series of personalizing technologies that guarantee a self, while continually being at risk of replacement in a seemingly never-ending dance. Yet, to the degree the Invisible Committee suggest a return to some notion of a fortified self by removing prosthesis, they miss a series of steps. The problem cannot be simply prosthesis but rather the way interactions with the outside are intensified because the instinct for destruction is linked to this moment of expulsion. The simple negation of prosthesis or of *technē* fails to hit the mark. We require alternately a moment that does not move to incorporation or expulsion—not management but something like a reaching out to objects that would avoid both identification and aversion to them.

In other words, the Invisible Committee continue to assume a discourse of mastery of the self over the self and not just the self's management of its various prostheses. As the reading of Foucault and Freud together suggests, it is mastery of the world by *technē* and the self by the test that fundamentally involves a relation to the negative; mastery as well as management are not affirmative in a way that might be able to block the move to incorporation, be it of the weak or strong variety. This raises the question of those other forms of relation to mastery that might forgo an intensification of biopower via mastery of the self. Perhaps we are speaking about apprenticeship as a form of nonmastery—to a self that is not driven solely by self-protection. Foucault recognizes such a perspective when he links governing, be it of the self or the polis, to a group as always enjoying a relation to power and mastery. This might well be the point at which governmentality fails to respond to biopower; indeed, it cannot, given the role that the negative plays therein.

Practicing *Bíos*

On the problem of the negative and responses to biopower, the question we need to ask ourselves is this: are there *technē* whose effects cannot be measured solely in mastery? To prepare the ground for an answer, we first need to acknowledge a problem with our lexicon while imagining other ways of referring to *technē* that would not evoke the negative or mastery. In limited fashion, I want to propose the notion of practice for reasons that have to do with the connections running between *practice* and the other key terms of this chapter, namely, *attention* and *play*. On this score, Pierre Bourdieu's insight into the relation of genealogy to practice precedes us. Speaking of maps and navigating abstract spaces, he writes,

> The gulf between this potential, abstract space, devoid of landmarks or any privileged centre—like genealogies, in which the ego is as unreal as the starting-point in a Cartesian space—and the practical space of journeys actually made, can be seen from the difficulty we have in recognizing familiar routes on a map or town-plan.[67]

To speak of a practice of *bíos* is to speak of the potential, abstract space of a genealogy whose origin is impossible to mark, given that where a self requires borders (so as to have coordinates and hence to navigate), *bíos* does not: in this sense, *bíos* and not the self is the privileged object of genealogy.[68] The task will be to locate the elements of a practice of *bíos* that do not fall headfirst into mastery, that do not immediately limit the potential of such a genealogical space of practice—in other words, an impolitical space that does not depend on representation since representation limits the possibilities precisely of attention as an impolitical practice. As such, a practice of *bíos* runs parallel with Foucault's view of critique as genealogical in that it "will not deduce from the form of what we are what it is impossible for us to do and to know; but it will separate out, from the contingency that has made us what we are, the possibility of no longer being, doing, or thinking what we are, do, or think."[69] *Bíos* and genealogy would therefore be inscribed within the horizon of practices by which one moves away from the self and its *technē*. To the degree that a practice of *bíos* (and not a *technē* of the self) would be one that does not move to reinforce the self and its extensions, it is "disconnected from the intensification of power relations," which Foucault saw at the heart of the notion of critique.[70]

Here I want to propose that we consider attention as the practice that will inform *bíos*. Let's immediately admit that not everyone agrees with such a perspective on attention as a practice worth practicing precisely because it appears unable to maintain any distance from intensifying power relations. In his seminal discussion of optical devices of the nineteenth century in *Techniques of the Observer*, Jonathan Crary, for instance, notes the thantopolitical weight of these apparatuses because they "involved the arrangements of bodies in space, regulations of activity, and the deployment of individual bodies." "They were," he goes on, "techniques for the management of attention, for imposing, homogeneity, anti-nomadic procedures that fixed and isolated the observer using partitioning and cellularity . . . in which the individual is reduced as a political force."[71] In a reading that will show how applied knowledge to the body increases attention "in the pure objectivity of perception," Crary will find mankind deeply enmeshed in the same web of *technē* of the self understood as mastery of the world that was the subject of the last pages of Foucault's critique in *Hermeneutics of the Subject*. In his 2002 follow-up, *Suspensions of Perception*, Crary draws a number of compelling conclusions from the earlier work to see attention as fundamentally problematic for modernity. "It is possible," he writes, "to see one crucial aspect of modernity as an ongoing criss of attentiveness, in which the changing configurations of capitalism continually push attention and distraction to new limits and thresholds, with an endless sequence of new products, sources of stimulation, and regulating perception."[72] For Crary, attention masks strategies of control and offers much less resistance to it (indeed, he will speak of attention as a phenomenon that drifts). Perhaps for this reason, Crary often inscribes attention in disciplinary horizons and, so doing, discounts the potential power of attention. But what if we inscribe attention in another horizon, of practice or play that attention provides? Might a reinforced attention offer more resistance to an intensification of power relations than what Crary suggests?

Others, however, have noted the stunning power of attention. Merleau-Ponty, for instance, in his beautiful essay "Attention and Judgment," remarks that "attention creates nothing, and it is a natural miracle . . . which strikes up like sparks those perceptions or ideas capable of providing an answer to questions that I was asking."[73] He suggests that "attention is therefore a general and unconditioned power in the sense that at any moment it can be applied indifferently to any content of consciousness. Being everywhere barren, nowhere can it have its own purposes to fill."[74] The power of attention for Merleau-Ponty differs from those powers that have "purposes to fill"—attention is less expedient than those other powers to the degree one can employ it whenever and wherever one wants.[75] For Merleau-Ponty, attention attends, which is to say not only that attention waits but also that it stretches toward objects while not taking possession of them (we can hear echoes of Freud's own perspective on judgment in the earlier passage in which he notes how judgment "affirms or disaffirms the possession by a thing of a particular attribute"). Key is the sense of stretching toward without taking hold of, of coming up to the object and waiting. Attention in this attending to runs counter to judgment. Writing of the nexus of intellectualism and judgment, he argues that "ordinary experience draws a clear distinction between sense experience and judgment. It sees judgment as the taking of a stand, as an effort to know something which shall be valid for myself every moment of my life, and equally for other actual or potential minds; sense experience, on the contrary, is taking appearance at face value. . . . This distinction disappears in intellectualism, because judgment is everywhere where pure sensation is not."[76] For Merleau-Ponty, attention moves toward possessing, which is to say, draws up to and simultaneously withdraws from judgment. On his read, attention offers a mode of approaching the object without incorporating or expelling.

What forms of life may be said to emerge from a practice of attention?[77] On one level, it is a form of life that avoids being captured by the self's acquisitive power, and here Merleau-Ponty's division of attention into two registers is helpful: secondary attention for him "would be limited to recalling knowledge already gained."[78] It is homologous to "acquisition"—to the process of memory linked to an identification with the subject of memory,[79] an acquisition that signals the process of incorporation and expulsion described earlier. Primary attention, however, does not merely "elucidate pre-existing data." It brings about "a new articulation . . . by taking them [data] as figures against a horizon. It is precisely the original structure which they introduce that brings out the identity of the object before and after the act of attention."[80] Obviously, we must not overlook the difficulties of Merleau-Ponty's reading of attention as bringing out the identity of the object both before and after, given such a perspective's dependence on the notion of horizon as providing an ultimate transcendence to the object/figure: the appearance of the figure against the horizon guarantees identity. And so we require ways of speaking about attention that would allow us to continue to hold open a space between figure and horizon, that do not merely postpone a decision about identity until after attention has run out. We require a practice that prolongs the time of attention so as to hold open the spatial interval between figure and horizon.

Haeccetic Attention

On this score, Gilles Deleuze and Félix Guattari's discussions of haecceity move to the center. In *A Thousand Plateaus,* they introduce the term *haecceity* to describe the kinds of possible spaces that call to mind the kind of space constructed by the attention I want to think here. They argue that we must think of haecceity as consisting not "simply of a decor or backdrop that situates subjects, or of appendages that hold things and people to the ground"; instead, haeccetic space encompasses "the entire assemblage that is defined by a longitude and a latitude, by speeds and affects, independently of forms and subject which belong to another plane."[81] In lieu of figure and horizon, they prefer to speak of composition. Thus "the street enters into composition with the horse, just as the dying rat enters into composition with the air, and the beast and the full moon enter into composition with each other." In their analysis, the composition between figures and between the former horizon and the figure does not lead immediately to a before and after in which attention has fixed the horizon and the figure but rather points to "the potentialities of becoming within each assemblage." For Deleuze and Guattari, "a haecceity has neither beginning nor end, origin nor destination: it is always in the middle."[82] Deleuze and Guattari's reading brings us closer to a form of attention that resides between form and horizon, form and world. So much so that a practice of attention able to meet the demands of biopower today is one that holds open a space in which potentialities for becoming are allowed to emerge. We note as well a not-so-subterranean connection in their discussion of the haeccetic space. A space of becoming is also a space in which elements move into composition with each other.

Deleuze details more features of such a compositional space in his reading of couples and triptychs in his study of Francis Bacon:

> The triptych is undoubtedly the form in which the following demand is posed most precisely: there must be a relationship between the separated

parts, but this relationship must be neither narrative nor logical. The triptych does not imply a progression, and it does not tell a story. Thus it too, in turn, has to incarnate a common fact for diverse Figures. It has to produce a "matter of fact."[83]

In Deleuze's formulation, art intersects with a "matter of factness" that relates different figures to each other, though not immediately to any kind of horizon. This matter of factness provides us with another way of thinking attention—a practice of attention would be one that allows a form of life to register the matter of factness of perception—that attends to sense perceptions in the same way that a triptych incarnates a common fact with no progression and with no move to bordering. The emphasis on incarnation in the passage, the fleshing out that takes place in haecceity as a "common" fact, is one we ought to take seriously as it also suggests a commonality between figures not joined to a defense of borders.

Consider too that attention and haecceity share a horizon with composition. Composition, as the *Oxford English Dictionary* tells us, comes from *com* (together or with) and *posere* (to place or to put down)—thus its meaning, "to make by putting together elements."[84] Composition strongly connotes a kind of relationality, with the emphasis not falling on the subject existing behind or before the act of picking up and putting down but on a kind of immanence that resides in the elements themselves. At the same time, we note a creative component to composition, which Deleuze discusses in *Cinema 1* in the context of the philosophy of montage. There he describes montage as a power that "is able to start fresh every instant, of starting afresh itself, and in this way confirming itself for itself, by putting the whole stake back into play each time."[85] The work of montage is the "work of a hand that touches, not of a hand that seizes," he will say.[86] This mode of nonseizing as a technique of modern art implies a model for attention as it, too, avoids seizing or judging.

A practice of attention equal to the demands of biopower would be one that mirrors such a perspective on montage. It is a part of what Félix Guattari refers to as a "new aesthetic processual paradigm."[87] For Guattari, the new aesthetic paradigm concerns an "aesthetic power of feeling" that "is equal in principle with the other powers of thinking philosophically."[88] What sets aesthetic power apart is not the reference to institutional art, or even its "dimension of creation in a nascent state, perpetually in advance of itself"; rather its privileged status involves a relation to "spheres of exteriority" that are not "radically separated from the interior. . . . There isn't really any exteriority."[89] The new paradigm's decisive feature is creation:

> The incessant clash of the movement of art against established boundaries . . . its propensity to renew its materials of expression and the ontological texture of the precepts and affects it promotes brings about if not a direct contamination of other domains then at the least a highlighting and a re-evaluation of the creative dimensions that traverse all of them. Patently, art does not have a monopoly on creation, but it takes its capacity to invent mutant coordinates to extremes: it engenders unprecedented, unforeseen and unthinkable qualities of being.[90]

For Guattari, the aesthetic paradigm implies a whole series of techniques that emphasize a "creative instance" rather than simply the thing created. In short, "one does not situate qualities or attributes as secondary in relation to being or substance; nor does one commence with being as a pure empty container (and a priori) of all the possible modalities of existing. . . . The emphasis is no longer placed on Being . . . it is placed on the manner of being."[91] Guattari intimates that a helpful contamination of attention with such an ontological view of art might help attention acquire some of the power of the creative—might move it toward a creative dimension and hence change its ontological registers.

What would a practice of creative attention look like? By moving toward the creative, such a practice would emphasize Being less and more modes of being, the manner of being of the objects of perception that appear in the relation that they share with other objects of attention—in the compositional and haeccetic space that attention shares with art.[92] Equally, such a practice of attention would not jump to marking the attributes of being or substance. We saw Heidegger enacting exactly this in the first chapter and Agamben in the second, when writing and life were respectively determined as proper and improper, depending on the quality attributed to each. An attention that holds together elements in a kind of compositional space does not posit a division between proper and improper but notes where they are located in such a space. It provides coordinates and fails to negate. Why call such a practice affirmative? Because it affirms what emerges out of the autopoeitic instances of the aesthetic paradigm. In this regard, Guattari writes that "the decisive threshold constituting this new aesthetic paradigm lies in the aptitude of these processes of creation to auto-affirm themselves as existential nuclei."[93] In other words, in the move toward attention as creation as ontology, the basic elements of existence are themselves affirmed. Following Guattari's reading of the aesthetic paradigm, attention as a practice might be seen as giving birth to immanent forms of life.

Our question must be how to potentialize attention because to do so would be a way of auto-affirming creation and, with it, an affirmative biopolitics. Surely one of the first ways will involve changing our conceptions of attention by bringing it closer to the realm of aesthetics. To do so, we need to emphasize attention's creative, compositional side—to see it as a practice of placing elements together from which forms of *bíos* may emerge not thought entirely or solely in terms of the self. Key in such a perspective on attention is the nonseizing that is at the heart of a compositional space. To compose, one must not seize; the act of putting together in fact precludes seizing the object—a grasping or clutching of one perception over another. Here I am reminded of Walter Benjamin's notion of immanent critique sketched in "The Concept of Art Criticism," in which he distinguishes between relationality and judging. This "completely different kind of criticism" is "not concerned with judging." "Its centre of gravity," he goes on, "lies not in the estimation of a single work but in demonstrating its relations to all other works, and ultimately to the idea of art."[94] A practice of attention as immanent critique would not seize by judging, incorporating, or expelling but would allow one to uncover the relationality inherent in the elements that make up the haeccetic space. A capacity is required to hold a perspective on the figures to which one is attending and, so doing, to shift the center of gravity to the relational side of the series of objects of perception. By holding open the interval between figure and horizon, object and meaning, a space emerges in which to attend to relationality.

The Metaphysics of Play

This holding apart of figure and horizon has another name, and that is "play." That we should consider play as a practice of attention, as a practice of *bíos,* is slightly unexpected when we remember how a number of twentieth-century philosophers looked to play for its antimetaphysical properties while ultimately discounting its potential.[95] On this score, George Bataille comes to mind. In the essay "Unknowing and Rebellion," he responds to a young medical intern who tells him that everything "comes down to the instinct of self-preservation." Bataille's response?

> My conception is surely less out of date. . . . It consists in saying that all is play, that being is play, that the idea of God is unwelcome and, furthermore, intolerable, in that God, being situated outside time, can be only play, but is harnessed by human thought to creation and to all the implications of creation, which go contrary to play (to the game).[96]

Bataille's task in the essay is to think "the possibility of a philosophy of play," which will lead him ultimately to substitute play with game, that is, "to think and be the game, make of the world and ourselves a game on condition that we look suffering and death in the face." He concludes soon after that "my thought has but one object, play, in which my thinking, the working of my thought dissolves."[97] Bataille's philosophy of play as involving the unknown and a rebellion against his own thought separates play from creation and hence surprisingly (for Bataille) is one that fails to take up the possibility of contaminating play with creation or, for that matter, attention with aesthetics.[98] Equally, it fails because Bataille sees play as part of a dialectic of the master and the slave: he who plays a game, the slave, so as to vanquish the master masters himself. The antinomy that Bataille recognizes but from which he cannot free himself is the negative place awarded death vis-à-vis the game. "The philosophy of play appears, in a manner that is fundamental, to be truth itself: common and indisputable: it is, nevertheless, out of kilter in that we suffer and die."[99] Essentially, then, Bataille thinks a form of play that would force the player to "look suffering and death in the face."[100] The advantage that attention as a creative form of play would have is that it does not attach a predicate to its objects. One does not play *at* something, namely, a game, but *with* something. In the case of attention as an affirmative practice, we might well say that one plays with the notion of self.[101]

Jacques Derrida, of course, is the other philosopher who, more than any other, takes up the philosophical merits of play. In "Structure, Sign, and Play in the Discourse of the Human Sciences," Derrida, writing with Lévi-Strauss in mind, places play "in tension" with history and then goes on to link play to presence:

> Play is the disruption of presence. The presence of an element is always a signifying and substituitive reference inscribed in a system of differences and the movement of a chain. Play is always play of absence and presence, but if it is to be thought radically, play must be conceived of before the

> alternative of presence and absence. Being must be conceived as presence or absence on the basis of the possibility of play and not the other way around.[102]

For Derrida, play precedes the oscillation between presence and absence. In the ultimate space of the play of difference, of difference as play, one can make out the effects of the contrary of play that will go by the name of Lévi-Strauss's ethic of innocence and the absent origin:

> If Levi-Strauss, better than any other, has brought to light the play of repetition and the repetition of play, one no less perceives in his work a sort of ethic of presence, an ethic of nostalgia for origins, an ethic of archaic and natural innocence, of a purity of presence and self-presence in speech—an ethic, nostalgia, and even remorse, which he often presents as the motivation of the ethnological project when he moves toward the archaic societies which are exemplary societies in his eyes.[103]

Lévi-Strauss's fractured present results in a negative thinking of play that continues to traffic in nostalgia and sadness. At the same time, Derrida adds, another form of play will be found that affirms the world, which is "the affirmation of a world of signs without fault, without truth, and without origin."[104] Such an affirmative form of play is problematic, however, for Derrida because it is still concerned with security and presence: "There is a *sure* play: that which is limited to the *substitution* of *given,* and *existing, present* pieces."[105] Ultimately, affirmation as an attribute of play is doomed because no such thing as presence can be said to exist without absence, and hence no ultimately secure mode of play is available. In other words, affirmation continues to be held hostage to knowing what is to come. Rather than affirmation, Derrida prefers "the *seminal* adventure of the trace," which is what arises when affirmation surrenders "to *genetic* indetermination."[106] Clearly affirmation for Derrida is problematic because as it turns away from the origin, affirmation "tries to pass beyond man and humanism, the name of man being the name of that being who, throughout the history of metaphysics or of ontotheology—in other words, throughout his entire history—has dreamed of full presence, the reassuring foundation, the origin and the end of play."[107] Presence kills off play.

Much remains to be said in this admittedly brief encapsulation of Derrida's perspective on play. Yet what stands out in a chapter on attention as play is the negative tone that characterizes Derrida's assertion of the trace's adventure, in particular, the weight he awards determination. The Nietzschean joyful affirmation of man gives way to "genetic indetermination." Derrida refuses, in my view mistakenly, to countenance something like the possibility of an affirmation not caught by sure forms of play. In this regard, it is helpful to compare Deleuze's reading of Spinoza's theory of negation with Derrida's. For Deleuze, that theory is based on "the difference between distinction, always positive, and negative determination: all determination is negative."[108] In Derrida's move to indetermination, one might well register a kind of remainder in the seminal adventure of the trace, something perhaps like a negative tonality. Deleuze, quoting Spinoza, notes,

> "Corresponding to positivity as infinite essence there is affirmation as necessary existence" (*Ethics*, I, 7 and 8). That is why all the attributes, which are really distinct precisely by virtue of their distinction without opposition, are at the same time affirmed of one and the same substance, whose essence and existence they express (I, 10, schol. 1 and 19).[109]

Deleuze's gloss of Spinoza offers a way forward from the negative tonality of indeterminacy by seeing play as a process of distinguishing without bringing in its wake oppositions and possessions of attributes—not presence as opposed to absence but instead an ontological force expressed by these attributes. To bring the discussion back, then, to practices, attention would name the practice that does not move to determining the objects of perception by calling forth their opposite—by judging them in Freud and Merleau-Ponty's perspective—but one that marks their distinctive attributes, that names them and, in so doing, distinguishes and affirms them. One does not find anything like the substituting of present pieces one for another but rather in a holding together of the elements of the compositional space, a possibility that other itineraries might become available. In other words, it might well be that attention itself is a form of unsure play.[110]

Aesthetics, Play, Creation

The conjunction of play and attention returns us to the notion of the creative that was central to my earlier gloss of Guattari's reading of the "aesthetic paradigm" and Derrida's notion of play. Play, lest we forget, is not simply about adventure—playing with something—but also includes a notion of creativity. To play often evokes something like a creative living element.[111] And so, what elements do attention and play share, that is, what would attention as play and play as attention look like? The answer will be found in a notion of immanent play that resembles a reaching out. To see what I mean, I want to introduce the notion of play within a context of writings about children at play. In particular, a number of pages from D. W. Winnicott's *Playing and Reality* come to mind, especially those in which he notes the importance of the interaction of a resting state and creation.[112] Winnicott thinks play not only in the handling of the object but also more broadly as emerging out of the interaction between a resting state, which he will gloss as a therapeutic setting, and what he calls "a [creative] reaching-out."[113] His meaning on this score is not always clear, but he appears to recall the earlier discussion of a practice of attention, and hence the move from resting to reaching out seemingly entails a creative attribute. Such a reaching out does not for Winnicott (as it does not for Deleuze in the earlier quote on montage or for Merleau-Ponty on judging) involve the moment of holding or grasping but appears as the mode by which a reaching out in play takes place for the child. He calls this between space "a space of relief" from the strain of relating inner and outer reality, "an intermediate area of experience which is not challenged," which he will then note as being "in direct continuity with the play area of the small child who is 'lost' in play."[114] The intermediate area, he tells us, "constitutes the greater part of an infant's experience" and throughout life is retained in the "intense experiencing that belongs to the arts and to religion and to imaginative living, and to creative scientific work."[115] Winnicott's child "lost" in play deserves a much longer discussion, what some call the "absorption model of play" thought in relation to those later moments of "imaginative living" that he calls "intense experiencing." What they share is not the feature of being lost or absorbed in play, however, but a shared origin in the mode of handling or reaching out. They also share another feature, one often overlooked, namely, that play is precarious as well as creative, that it is "always on the theoretical line between the subjective and that which is objectively perceived," which may come to an end at any instant. Indeed, for Winnicott, the precariousness of play precisely makes it a "creative experience." Thus he will call play an "experience in the space-time continuum, a basic form of living." It is for this reason that playing as a creative reaching out for Winnicott is itself "a therapy."[116]

It bears repeating that Winnicott's view of play as premised on a creative reaching out in opposition to a noncreative seizing echoes the roots of the word *attention* as "a sense of tension, of being stretched, and also of waiting."[117] In the reaching out of attention, we have a form of play that begins to enjoy a relation to a basic form of living, an imaginative living that is able to provide some relief from "the strain of inner and outer reality"—in other words, from the continual move to introject or expel that lies at the heart of the instinct for self-preservation and for destruction. Yet we can go further in linking creative attention and *bíos* through the notion of play. Play and attention also share how they withdraw from possessing, a mode that does not immediately make the object of perception or the toy one's own. Play may become the ground for other forms of life or modes of being to arise, in which the supposed content of *bíos* as self will determine how it is that one plays at life. Here it may useful to turn to those wonderful pages from Walter Benjamin's "The Cultural History of Toys," in which Benjamin gives a philosophical classification of toys:

> As long as the realm of toys was dominated by a dour naturalism, there were no prospects of drawing attention to the true face of a child at play. Today we may perhaps hope that it will be possible to overcome the basic error—namely, the assumption that the imaginative content of a child's toys is what determines his playing; whereas in reality the opposite is true. A child wants to pull something, and so he becomes a horse . . . he wants to hide, and so he turns himself into a robber or a policeman. . . . Because the more appealing toys are . . . the further they are from genuine playthings; the more they are based on imitation, the further away they lead us from real, living play.[118]

Benjamin opposes a "real, living play" to a dour, naturalistic play thought in terms of the greater appeal of toys. Real living play emerges, instead, in relation to how the child is able to transform herself, to become the plaything. The plaything, we will want to say, leads to becoming, to being absorbed, to borrow the language with which play is often described (the other being slippage).[119] For Benjamin, real play is premised on the presence of playthings that allow the player to be transformed. In different words, behind the genuine plaything lies the self's shadow, and one does not play thanks to the features of the toys but rather on account of their lack of imaginative content. Benjamin's perspective allows us to see the shared capacity of attention and play to forgo *imitation,* which is yet another term for describing a particularly insidious form of incorporation and expulsion.

How might we extend this notion of play into the realm of biopolitics as a form of resistance to biopower? Can we play at forms of life not captured by the self? Can one modify the sense of self as one modifies a toy in the act of playing so as to create a broader perspective on *bíos,* on those modes of being so unartfully captured by the self? On this score, let's consider another passage from Benjamin: "But we must not forget that the most enduring modifications in toys are never the work of adults, whether they be educators, manufacturers, or writers, but are the result of children at play."[120] Looking to children at play, perhaps we can make out the horizon of a living art in play that would allow us to see how the self, by withdrawing to play as *bíos,* might be able, if not to block, then to slow down the speed with which borders and defenses and, with them, the instinct for destruction are made manifest. A *technē* of *bíos* thought through play might be one yet unexplored way to forgo "the dour naturalism" of biopolitics today, in which the object of politics would be merely biological life or that would have the object of life be thinkable only as part of a negative politics.

Creatures, Contradictory and Virtuous

After tracing the links between *technē* and thanatos, what are we left with finally? Framing the question in this way is already helpful: the question is less what is to be done and more what kind of self remains for a practice of *bíos.* Writing near the end of the twentieth century, William Connelly speaks of slackening in a context of discipline that may be useful:

> Since the self is not "designed" to fit perfectly into any way of life, we must anticipate that every good way of life will both realize something in the self and encounter elements in the self resistant to its form; and we should endorse the idea of slack as part of our conception of the good life. An order with slack can sustain itself well without the need to organize the self so completely into a creature of virtue. For the more an order needs virtue the more it eventually authorizes the extension of disciplinary strategies to secure it.[121]

For Connolly, slackening the self will reinscribe it in horizons different from mere virtue. Yet the stakes, in my view, move well beyond discipline and virtue to the overarching reading that a number of contemporary philosophers have given of biopower at the crossroads of technology and death. Another way of putting it might be, rather than speaking of the good life, we ought instead to focus on forms of *bíos* able to loosen the self's relation to a mastery of the world. The creature of virtue produced by the test gives way to a being capable of playing with the notion of the self in different contexts and situations. Doing so, the self becomes less available to the interpellations of institutions by performing less masterfully. The move from the bordered self to the slackened subject of the practices of *bíos* minimizes the contact that borders inevitably share with thanatos. Let's also note that the object of a practice of *bíos* would not necessarily be virtuous either. If there is a question of virtue, it concerns the virtue of the nonvirtue of *bíos*—a virtue in seeing the self as too limited and limiting.

One surely could respond that a practice of *bíos* thought through attention and play seems too abstract—that it fails to provide a decent wage or allow one to spend more time with one's children. In other words, what does a practice of *bíos* offer the subject materially right now? My own impression is that we ought to avoid any rushes to judgment about practices as opposed to resistance—about improper political practices as opposed to proper political techniques of rebellion, which would simply reenact some of the same difficulties I have sketched here around proper and improper forms of life and *technē*. On this score, consider those pages from Nietzsche in which he puts forward the case for a relaxing of borders around perspectives linked merely to the self:

> A point of view: in all valuations there is a definite purpose: the *preservation* of an individual, a community, a race, a state, a church, a belief, or a culture.—Thanks to the fact that people *forget* that all valuing has a purpose, one and the same man may swarm with a host of contradictory valuations, and *therefore with a host of contradictory impulses*. This is the *expression of disease in man* as opposed to the health of animals, in which all the instincts answer certain definite purposes. This creature full of contradictions [*dies widerspruchsvolle Geschöpf*], however, has in his being a grand method of acquiring knowledge: he feels the pros and cons, he elevates himself to *Justice*—that is to say, to the ascertaining of principles *beyond the valuations of good and evil*. The wisest man would thus be gifted with mental antennae [*der gleichsam Tastorgane*] wherewith he could understand all kinds of men, and with it he would have his great moments, when all the chords of his being would ring in *splendid unison*. . . . A sort of planetary movement [*eine Art planetarischer Bewegung*].[122]

Perspectives police borders. They preserve and protect. The cost of such a valuing and evaluating, however, is to set up borders around a self that will then continue to require defending. Nietzsche's solution is not, we should note, to point a way to a return to the health of animals but rather to embrace the contradictions of the diseased human being.[123] We should also be clear that this is not Connelly's creature of virtue either, but a different one, a creature of contradictions, "dies widerspruchsvolle Geschöpf," and one aware of these contradictions. Perhaps we might find a space here for inserting the practices of attention and play as modes of registering the contradictory features of Nietzsche's creature, not as a return to the animal but rather to the being who receives the gift that makes him wise to the degree other creatures—other forms of *bíos*—come into view. In this sense, we might see the practices of *bíos* through attention and play as serving definite purposes, namely, moving the self toward greater openness and relationality as opposed to defense. Here a convergence appears between these practices I have discussed and the result of the contradictions, namely, a mode of living in which the wisest "feels the pros and cons" yet does not know them in a traditional sense. Acknowledging the different perspectives of the contradictory creature, the subject of attention and play moves beyond good and evil, beyond true and false in Foucault's terms, indeed beyond mastery of the self by the self. Holding open the space for attention creates conditions in which an "understanding of all men" results—what we should not hesitate to inscribe in a political register as "a planetary movement." Note, too, that Nietzsche suggests another possibility for resisting the complicity of the self with biopower through mastery, those moments when "biopower takes over the activity of care of the self."[124] These will include modes of ascertaining, elevating, and feeling that do not necessarily or immediately involve "a care of the self."[125] Admittedly, the task of resisting care and, with it, mastery of the self is not an easy one, but that is where the value added of a practice of *bíos* through attention and play might be felt most: in helping us create a breach between care and mastery, between a care for the self felt first in terms of forms of *bíos* and known only after in terms of mastery.[126]

NOTES

Preface

1 Following the work of Heidegger and Foucault, *technē* refers to craft or art and has decisively fewer negative connotations than does *technology.* For Greek usage, see Aristotle, *Nicomathean Ethics,* chapter 6, as well as Book VI of Plato's *Republic.*

2 Giorgio Agamben, *Stanzas: Word and Phantasm in Western Culture,* trans. Ronald L. Martinez (Minneapolis: University of Minnesota Press, 1993), esp. 148; Agamben, *The Coming Community,* trans. Michael Hardt (Minneapolis: University of Minnesota Press, 1993); Agamben, *Means without End: Notes on Politics,* trans. Vincenzo Binetti and Cesare Casarino (Minneapolis: University of Minnesota Press, 2000).

3 Roberto Esposito, *Communitas: The Origin and Destiny of Community,* trans. Timothy Campbell (Palo Alto, Calif.: Stanford University Press, 2009).

4 Martin Heidegger, *Parmenides,* trans. André Schuwer and Richard Rojcewicz (Bloomington: Indiana University Press, 1992).

5 Giorgio Agamben, *The Kingdom and the Glory: For a Theological Genealogy of Economy and Government,* trans. Lorenzo Chiesa and Matteo Mandarini (Palo Alto, Calif.: Stanford University Press, 2011); Agamben, *The Sacrament of Language: An Archaeology of the Oath,* trans. Adam Kotski (Palo Alto, Calif.: Stanford University Press, 2010); and Agamben, *What Is an Apparatus?* trans. David Kishik and Stefan Pedatella (Palo Alto, Calif.: Stanford University Press, 2009).

6 Roberto Esposito, *Terza persona: politica della vita e filosofia dell'impersonale* (Turin, Italy: Einaudi, 2006); Esposito, *Termini della politica* (Milan, Italy: Mimesis, 2008); Esposito, "The Dispositif of the Person," *Law, Culture, Humanities* 7, no. 3, forthcoming.

7 Peter Sloterdijk, *Rage and Time: A Psychopolitical Investigation,* trans. Mario Wenning (New York: Columbia University Press, 2009), and "Rules for the Human Zoo: A Response to the *Letter on Humanism,*" *Environment and Planning D: Society and Space* 27 (2009): 12–28.

8 Michel Foucault, "What Is Enlightenment?," in *Ethics: Subjectivity and Truth,* ed. Paul Rabinow, trans. Robert Hurley et al. (New York: New Press, 1997), 319.

9 Michel Foucault, *Security, Territory, Population: Lectures at the Collège de France 1977–1978,* trans. Graham Burchell (New York: Picador, 2009), and Foucault, *The*

Hermeneutics of the Subject: Lectures at the Collège de France 1981–1982, trans. Graham Burchell (New York: Palgrave, 2005).

1. Divisions of the Proper

1 An abbreviated list would include David R. Cerbone, *Heidegger: A Guide for the Perplexed* (New York: Continuum, 2008); Jeff Malpas, *Heidegger's Topology: Being, Place, World* (Cambridge, Mass.: MIT Press, 2006); Ruth Irwin's compelling *Heidegger, Politics, and Climate* (New York: Continuum, 2009); Matthew Calarco, *Zoographies: The Question of the Animal from Heidegger to Derrida* (New York: Columbia University Press, 2008); Richard Rojcewicz, *Gods and Technology: A Reading of Heidegger* (Albany: SUNY Press, 2006), esp. part II; and Alan Milchman and Alan Rosenberg, eds., *Foucault and Heidegger: Critical Encounters* (Minneapolis: University of Minnesota Press, 2003).

2 Certainly the writer who comes closest is Jacques Derrida. Apropos of the hand in Heidegger, he writes: "The hand is monstrasity [*monstroisté*], the proper of man as the being of monstration. This distinguishes him from every other *Geschlecht,* and above all from the ape." "Geschlecht II," in *Deconstruction and Philosophy: The Texts of Jacques Derrida,* ed. John Sallis (Chicago: University of Chicago Press, 1987), 169. Equally, Derrida's reading of the typewriter informs a large part of my reading of propriety in Heidegger's thought as well as the threat of improper writing. Another passage makes this clear: "Finally, the typewriter would dissimulate the very essence of the writing gesture and of writing. . . . This dissimulation or this veiling is also a movement of withdrawal or subtraction. . . . And if in this withdrawal [*retrait*] the typewriter becomes '*zeichenlos,*' without sign, unsignifying, a-signifying, that is because it loses the hand; in any case it threatens what in the hand holds speech safe [*garde la parole*] or holds safe for speech the relation of Being to man and of man to beings" (179). Compare as well Derrida's reading of the relation of the human to "as such" in *The Animal That Therefore I Am,* trans. David Wills (New York: Fordham University Press, 2008), 141–60. For a recent attempt to locate a biopolitical Heidegger as well as a biopolitical Derrida, see Gregg Lambert's contribution to the conference "Biopolitics and Its Vicissitudes" (Amherst College, April 17–18, 2009) titled "Globalatinization."

3 Heidegger, *Parmenides*; Heidegger, *Elucidations of Hölderlin's Poetry,* trans. Keith Hoeller (New York: Humanity Books, 2000); Heidegger, "The Question Concerning Technology," in *The Question Concerning Technology and Other Essays,* trans. William Lovitt, 3–35 (New York: Harper and Row, 1977); and Heidegger, "Letter on Humanism," in *Basic Writings from Being and Time (1927) to The Task of Thinking (1964),* ed. David Farrell Krell, 193–242 (San Francisco: HarperSanFrancisco, 1977).

4 And biopower. See Maurizio Lazzarato, "From Biopower to Biopolitics," http://www.generation-online.org/c/fcbiopolitics.htm (accessed June 11, 2011).

5 "Praxis and political reflection are operating today exclusively within the dialectic of proper and improper—a dialectic in which either the improper extends its own rule everywhere, thanks to an unrestrainable will to falsification and consumption (as it happens in industrialized democracies), or the proper demands the exclusion of any impropriety (as it happens in integralist and totalitarian states)." Agamben, *Means without End,* 116.7.

6 Roberto Esposito, *Bíos: Biopolitics and Philosophy,* trans. Timothy Campbell (Minneapolis: University of Minnesota Press, 2008); Esposito, *L'origine della politica: Simone Weil o Hannah Arendt?* (Rome, Italy: Donzelli, 1996); and Esposito, *Terza persona.*

7 On Heidegger and the tragic, see Jason Powell, *Heidegger's Contributions to Philosophy* (New York: Continuum, 2007), esp. 92–102. For quite a different reading of Heidegger and the tragic, see Gianni Vattimo, "Optimistic Nihilism," in *Common Knowledge* 6, no. 3 (1992): 43: "It is not a matter of persuading oneself that nihilism is good as it is; resignation must be accompanied by distortion. *Verwindung* begins with acceptance, with a change in the valence of nihilism; but it must be reinterpreted so that we postmoderns grasp the chance offered by our weakening sense of reality."

8 See, in particular, Friedrich Kittler, who employs Heidegger as a kind of ultimate authorization for his own theorization of discourse networks. Kittler, *Discourse Networks 1800/1900,* trans. Michael Metteer, with Chris Cullens (Palo Alto, Calif.: Stanford University Press, 1990).

9 Heidegger, *Parmenides,* 80.

10 Ibid. This evokes in its structure that other moment of Heidegger's dogmatism that Derrida traces: "Here in effect occurs a sentence that at bottom seems to me Heidegger's most significant, symptomatic, and seriously dogmatic. . . . The sentence in sum comes down to distinguishing the human *Geschlecht,* our *Geschlecht,* and the animal *Geschlecht,* called 'animal' . . . : 'Apes, *for example* [emphasis added], have organs that can grasp, but they have no hand.'" Derrida, "Geschlecht II," 173.

11 Ibid., 80. Compare Bernard Stiegler's reading of the historical in Heidegger: "The ambiguity of the invention of the human, that which holds together the *who* and the *what,* binding them while keeping them apart, is différance undermining the authentic/inauthentic divide." Stiegler, *Technics and Time: The Fault of Epimetheus,* trans. Richard Beardsworth and George Collins (Palo Alto, Calif.: Stanford University Press, 1998), 141.

12 On the importance of the difference between cause in Greek and Latin, see William Lovitt, "A Gespräch with Heidegger on Technology," *Man and World* 6, no. 1 (1973): 44–62.

13 Theodor Adorno, *The Jargon of Authenticity,* trans. Knut Tarnowski and Frederic Will (London: Routledge, 2002).

14 Heidegger, *Parmenides,* 80–81.

15 Ibid., 85.

16 André Schuwer and Richard Rojcewicz, "Translators' Foreword," in Heidegger, *Parmenides,* xiii. See, as well, Derrida's reflections on the famous photograph of Heidegger: "Since then I have studied all the published photographs of Heidegger, especially in an album bought at Freiburg when I had given a lecture there on Heidegger in 1979. The play and the theater of hands in that album would merit a whole seminar. If I did not forgo that, I would stress the deliberately craftsman-like staging of the hand play, of the monstration and demonstration that is exhibited there, whether it be a matter of the handling [*maintenance*] of the pen, of the maneuver of the cane that shows rather than supports, or of the water bucket near the fountain." Derrida, "Geschlecht II," 169.

17 Friedrich Kittler, *Gramophone, Film, Typewriter,* trans. Geoffrey Winthrop-Young and Michael Wurz (Palo Alto, Calif.: Stanford University Press, 1999); Timothy Campbell, *Wireless Writing in the Age of Marconi* (Minneapolis: University of Minnesota Press, 2006).

18 If we were to choose to follow Heidegger's thought on this point further, we would find that here, as elsewhere, Heidegger reproduces in the relation between proper and improper what Roberto Esposito describes as "a subtle ethical unevenness between the two terms that produces that dialectic of loss and re-discovery . . . of bewilderment and reappropriation." Esposito, *Communitas,* 98. Here we are far removed from the other more radical moments of Heidegger's thought, in which Dasein itself is understood "in its most proper impropriety." Ibid.

19 See esp. Esposito's introduction to *Communitas.* Compare in this regard Derrida's reflections on the relation of the hand and the gift: "This passage from the transitive gift, if such can be said, to the gift of what gives *itself,* which gives itself as being-able-to-give, which gives the gift, this passage from the hand that gives to the hand that gives *itself* is evidently decisive." Derrida, "Geschlecht II," 175. Although Derrida refuses to speak of this in biopolitical terms, there does emerge an implicit acknowledgment of the difference between a human giving and animal taking because the "*animal rationale* can only *take hold of, grasp, lay hands on the thing.*" Ibid.

20 Compare in this regard Simone Weil on idolatry and the personal: "The collectivity is not only alien to the sacred, but it deludes us with a false imitation of it. Idolatry is the name of the error which attributes a sacred character to the collectivity; and it is the commonest of crimes, at all times, at all places." Weil, "Human Personality," in *Simone Weil: An Anthology,* ed. Siân Miles (New York: Grove Press, 1986), 56.

21 Heidegger, *Parmenides,* 86.

22 Ibid.

23 Ibid., 86–87.

24 "When people had become acquainted with thermodynamics, and asked themselves how their machine was going to pay for itself, they left themselves out. They regarded the machine as the master regards the slave—the machine is there, somewhere else, and it works. They were forgetting only one thing, that it was they who had signed the order form." Jacques Lacan, "The Circuit," in *The Seminar of Jacques Lacan,* vol. 2, *The Ego in Freud's Theory and in the Technique of Psychoanalysis, 1954–1955* (New York: W. W. Norton, 1991), 83.

25 Donna Haraway, "Biopolitics and Postmodern Bodies," in *Symians, Cyborgs, and Women: The Reinvention of Nature,* 203–30 (New York: Routledge, 1991), and Hal Foster, *Prosthetic Gods* (Cambridge, Mass.: MIT Press, 2006).

26 Heidegger, *Parmenides,* 86.

27 Heidegger, "Letter on Humanism," 200.

28 Heidegger, *Parmenides,* 86.

29 Ibid.

30 Ibid.

31 Ibid. See Heidegger's *Introduction to Metaphysics* for further details on the relation of the oblivion of Being to nihilism: "But where is the real nihilism at work? Where one clings to current beings and believes it is enough to take beings, as before, just

as the beings they are. . . . Merely to chase after beings in the midst of the oblivion of Being—that is nihilism." Heidegger, *Introduction to Metaphysics,* trans. Gregory Fried and Richard Polt (New Haven, Conn.: Yale University Press, 2000), 217.

32 Ibid.

33 See Friedrich Kittler, "Media Wars: Trenches, Lightning, Stars," in *Literature, Media, Information Systems,* 117–29 (Amsterdam: OPA, 1997).

34 For a definition of the impolitical, see Massimo Cacciari, *The Impolitical: On the Radical Critique of Political Reason,* trans. Massimo Verdicchio (New York: Fordham University Press, 2009), esp. "Nietzsche and the Impolitical," 92–103, as well as Roberto Esposito's introduction to *Oltre la politica: Antologia del pensiero 'impolitico'* (Milan, Italy: Bruno Mondadori, 1996), 10: "The language of power is the only language of reality in the specific sense that it is incapable of speaking languages different from that of power."

35 Heidegger, "Question Concerning Technology," 16.

36 See the aforementioned works by Friedrich Kittler as well as Siegfried Zielinksi, *Deep Time of the Media: Toward an Archaeology of Hearing and Seeing by Technical Means,* trans. Gloria Custance (Cambridge, Mass.: MIT Press, 2008).

37 Heidegger, "Question Concerning Technology," 17.

38 Ibid., 19.

39 Ibid., 27.

40 Ibid., 26 (emphasis original).

41 Ibid., 27.

42 Ibid.

43 Ibid., 28.

44 Ibid., 26.

45 "The 'radical novelty' of modernity consists in the fact that 'now for the first time,' there is a distance, an alienation, that encourages and makes plausible the Archimedean project of utterly transforming the conditions of human existence." Dana R. Villa, *Arendt and Heidegger: The Fate of the Political* (Princeton, N.J.: Princeton University Press, 1996), 193.

46 Heidegger, "Question Concerning Technology," 33.

47 Ibid., 25.

48 Ibid., 27.

49 Heidegger, *Elucidations of Hölderlin's Poetry,* 42.

50 Ibid., 43. Philippe Lacoue-Labarthe takes up these issues in the recently translated *Heidegger and the Politics of Poetry,* trans. Jeff Fort (Urbana: University of Illinois Press, 2007).

51 The titles are too numerous to mention but would surely include Jacques Derrida, *The Post Card: From Socrates to Freud and Beyond,* trans. Alan Bass (Chicago: University of Chicago Press, 1987), as well as *Archive Fever: A Freudian Impression,* trans. Eric Prenowitz (Chicago: University of Chicago Press, 1996); Avital Ronell, *The Telephone Book: Technology, Schizophrenia, Electric Speech* (Lincoln: University of Nebraska Press, 1987); and Briankle Chang, *Deconstructing Communication: Representation, Subject, and Economies of Knowledge* (Minneapolis: University of Minnesota Press, 1995); and *Reading Matters: Narrative in the New Media Ecology,* ed. Joseph Tabbi and Michael Wutz (Ithaca, N.Y.: Cornell University Press, 1997).

52 Heidegger, *Elucidations of Hölderlin's Poetry,* 42.
53 Ibid.
54 For the relation of *Ge-stell* and mystery, see Véronique M. Fóti, *Heidegger and the Poets: Poiēsis/Sophia/Technē* (London: Humanities Press, 1992), 13: "Since *poiēsis* remains mindful of the 'mystery' or the lack of any inherent, positable reality in the happening of manifestation, it is able to awaken awareness of the destinal-historical ambiguity between totalizing and differential modalities of unconcealment in the *essence* of *Ge-stell.*"
55 Heidegger, *Elucidations of Hölderlin's Poetry,* 43.
56 Ibid., 47.
57 Heidegger, "Question Concerning Technology," 33.
58 Sloterdijk, "Rules for the Human Zoo."
59 Heidegger, "Question Concerning Technology," 34.
60 Esposito, *Communitas,* 99. Compare in this regard Heidegger's own defense of Nazism in *Introduction to Metaphysics,* 213: "In particular, what is peddled about nowadays as the philosophy of National Socialism, but which has not the least to do with the inner truth and greatness of this movement [namely, the encounter between global technology and modern humanity], is fishing in these troubled waters of 'values' and 'totalities.'"
61 Agamben, *Means without End,* 31.
62 The case of the radio is especially interesting in a context of improper writing, growing as it does out of wireless telegraphy and the transcription of dots and dashes into a written (and at times typed) message. What they share, in any case, beyond their impropriety in Heidegger's terminology is their positioning of the operator as "ordered," i.e., as one ordered as "standing reserve" and equally as one who requires saving.
63 See http://www.youtube.com/watch?v=MKTw3_PmKtc (accessed June 10, 2011).
64 Ernst Nolte, *Three Faces of Fascism: Action Francaise, Italian Fascism, National Socialism,* trans. Leila Vennewitz (New York: New American Library, 1969), 554.
65 As we will have occasion to note when discussing Sloterdijk's own brand of the thanatopolitical, the fact that Heidegger chooses to respond to Jean-Paul Sartre's question in the form of a letter is one that deserves our attention. On the relation of the philosophical to the letter in Sloterdijk, see Paul Rabinow, *Anthropos Today: Reflections on Modern Equipment* (Princeton, N.J.: Princeton University Press, 2003), 81–83.
66 Heidegger, "Letter on Humanism," 197.
67 Ibid.
68 See Roberto Esposito's "Prefazione" to *Categorie dell'impolitico,* vii–xxxii (Bologna, Italy: Il Mulino, 1999).
69 Heidegger, "Letter on Humanism," 197.
70 Ibid., 198.
71 Ibid., 199.
72 Ibid., 200.
73 See in this regard Malpas, *Heidegger's Topology,* esp. chapter 4.
74 See David Farrell Krell's helpful note with regard to care. Care is "a name for the structural whole of existence in all its modes and for the broadest and most basic

possibilities of discovery and disclosure of self and world. . . . 'Care' is the all-inclusive name for my concern for other people, preoccupations with things, and awareness of my proper Being. It expresses the moment of my life out of a past, into a future, through the present." Krell, *Basic Writings from Being and Time*, 199–200.

75 Heidegger, "Letter on Humanism," 203–4.

76 Compare in this regard Brett Buchanan's chapter "Disruptive Behavior: Heidegger and the Captured Animal," in *Onto-ethologies: The Animal Environments of Uexküll, Heidegger, Merleau-Ponty, and Deleuze*, 65–112 (Albany: SUNY Press, 2008).

77 Heidegger, *Parmenides*, 84.

78 Heidegger, "Question Concerning Technology," 27.

79 Heidegger, *Parmenides*, 84.

2. The *Dispositifs* of Thanatopolitics

1 Agamben, *Stanzas*; Agamben, *The Kingdom and the Glory*; Agamben, *The Signature of All Things: On Method*, trans. Luca D'Isanto, with Kevin Attell (New York: Zone Books, 2009); Agamben, *Sacrament of Language*.

2 Giorgio Agamben, *The Open: Man and Animal*, trans. Kevin Attell (Palo Alto, Calif.: Stanford University Press, 2004), 63–70.

3 "His [Agamben's] cultural universe, which stretched from Heidegger to Benjamin, from the dialectic of mysticism to Deleuzian immanence, is too complex to speak of influences." Antonio Negri, "Giorgio Agamben: The Discreet Taste of the Dialectic," in *Giorgio Agamben: Sovereignty and Life*, ed. Matthew Calarco and Steven DeCaroli (Palo Alto, Calif.: Stanford University Press, 2007), 120.

4 I have chosen not to translate *dispositif* for reasons that concern principally the associations the term loses should it be translated as "apparatus," as the recent English translation of the essay does. Principally, these are "the arrangement of troops in preparation for a military operation as well as their distribution, allocation, destination" and "the power of disposing of" ("disposition," *Oxford English Dictionary*).

5 Agamben, *Means without End*, 1.

6 Ibid., 116.

7 To some degree, all of Agamben's work may be inscribed in what amounts to a theory of the improper, first instituted in the chapter of *Stanzas* titled "The Proper and the Improper" and those important pages he dedicates to the relation of metaphor and the improper. "Metaphor, as the paradigm of signifying by improper terms (and according to baroque theorists, both the emblem and the impresa fall under this framework), becomes thus the principle of a universal dissociation of each thing from its own form, of every signifier from its own signified" (142). Agamben will later elaborate the "origins" of the desubjectification of the subject through the baroque improper since "in emblems . . . each word to its own signified is radically called into question" (142). See also in this regard Agamben's introduction to Giorgio Manganelli's *Contributo critico allo studio delle dottrine politiche del '600 italiano* (Macerata, Italy: Quodlibit, 1999), 7–18.

8 Compare on this relation Roberto Esposito's recent "Totalitarianism or Biopolitics:

Concerning a Philosophical Interpretation of the 20th Century," *Critical Inquiry* 34, no. 4 (2008): 633–45.

9 Agamben, *Stanzas,* 142. On the notion of second-order observation, see Niklas Luhmann, "Deconstruction as Second-Order Observing," *New Literary History* 24, no. 4 (1993): 763–82. For a brilliant reconstruction in terms of animal life, see the conclusion of Cary Wolfe, *Animal Rites: American Culture, the Discourse of Species, and Philosophical Theory* (Chicago: University of Chicago Press, 2003).

10 "To have mistaken such a naked life separate from its form, in its abjection, for a superior principle—sovereignty or the sacred—is the limit of Bataille's thought, which makes it useless to us." Agamben, *Means without End,* 7. See also Giorgio Agamben, *Homo Sacer: Sovereign Power and Bare Life,* trans. Daniel Heller-Roazen (Palo Alto, Calif.: Stanford University Press, 1998), 112–14.

11 "If this is the case, if the essence of the camp consists in the materialization of the state of exception and in the consequent creation of a space for naked life as such, we will then have to admit to facing a camp virtually every time that such a structure is created, regardless of the nature of the crimes committed in it and regardless of the denomination and specific topography it might have." Agamben, *Means without End,* 41–42. Yet, at the same time, this would presumably be the moment of the greatest potentiality as well: "The living being, which exists in the mode of potentiality [*potenza*] can its own proper impotence and only in this way possess its own proper potentiality [*potenza*]." Agamben, "La potenza del pensiero," in *La potenza del pensiero: Saggi e conferenza* (Vicenza, Italy: Neri Pozza, 2005), 281. Compare as well the brief passage that concludes "Idea of Power" in *L'idea della prosa*: "In work as in pleasure, mankind ultimately enjoys his own proper impotence." Agamben, *Idea della prosa* (Macerata, Italy: Quodlibit, 2002), 52.

12 Giorgio Agamben, *Remnants of Auschwitz,* trans. Daniel Heller-Roazen (New York: Zone Books, 1999).

13 Giorgio Agamben, *State of Exception,* trans. Kevin Attell (Chicago: University of Chicago Press, 2005), esp. 52–60.

14 Agamben, *Coming Community,* 62.

15 Ibid., 63–64.

16 Ibid., 65.

17 Ibid.

18 Agamben, *Remnants of Auschwitz,* 156.

19 Ibid., 155.

20 Ibid., 146–47.

21 Heidegger, "Letter on Humanism," 197.

22 Proof of which will be found in Slavoj Žižek's op-ed piece "Knights of the Living Dead," *New York Times,* March 27, 2007.

23 Agamben, *Remnants of Auschwitz,* 139.

24 Ibid.

25 See Milchman and Rosenberg, *Foucault and Heidegger,* esp. Hubert L. Dreyfus, "Being and Power Revisited," 30–54.

26 Agamben, *Remnants of Auschwitz,* 151.

27 "'Messianic and weak' is therefore that potentiality of saying which . . . also exceeds the act of saying itself. . . . This is the remnant of potentiality that is not

consumed in the act, but is conserved in it each time and dwells there. . . . It acts in its own weakness." Giorgio Agamben, *The Time That Remains: A Commentary on the Letter to the Romans,* trans. Patricia Dailey (Palo Alto, Calif.: Stanford University Press, 2005), 137.

28 Agamben, *Remnants of Auschwitz,* 75–76.

29 Ibid., 76.

30 Ibid., 83.

31 Ibid., 84.

32 Agamben, *Means without End,* 31.

33 Thus *people* as symbolized by the Jews is another name for "that naked life that modernity necessarily creates within itself but whose presence it is no longer able to tolerate in any way." *Ibid.,* 34. To read modernity as a question of tolerance for a mere form of life is to mark *zoē* as the thanatopolitical supposition of a political lexicon that still speaks of sovereignty, freedom, and democracy as if the division between *bíos* and *zoē* was nowhere to be found—as if modernity does not operate as some kind of machine, anthropological or otherwise, that limits tolerance to what it can systemically include and exclude. See in this regard Massimo Cacciari's recent essay on antiquity and tolerance in *La maschera della toleranza* (Milan, Italy: Bur-Rizzoli, 2006).

34 Agamben, *Remnants of Auschwitz,* 161. Compare on this point Agamben's more recent reading of archaeology and the *arché*: "The *arché* is not a given, a substance or an event, but instead a field of historical currents stretched between the anthropogenesis and the present, an ultra-history and history. As such . . . it can eventually render historical phenomena intelligible." Agamben, *Sacrament of Language,* 11.

35 Ibid.

36 Ibid., 162.

37 Agamben, *Signature of All Things,* 32.

38 Ibid., 31.

39 See Catherine Mills, *The Philosophy of Giorgio Agamben* (Montreal: McGill-Queen's University Press, 2009); Andrew Norris, ed., *Giorgio Agamben and the Politics of the Living Dead* (Durham, N.C.: Duke University Press, 2005); Leland de la Durantaye, *Giorgio Agamben: A Critical Introduction* (Palo Alto, Calif.: Stanford University Press, 2009); as well as Jeffrey S. Librett's "From the Sacrifice of the Letter to the Voice of Testimony," *diacritics* 37, nos. 2–3 (2007): 11–33.

40 Agamben, *Means without End,* 41.

41 Ibid., 43.

42 Agamben, *Homo Sacer,* 115.

43 Ibid., 114. This is not the first time Agamben extends a category of man far and wide. Writing in *Infancy and History,* Agamben traces a line of thought from Benjamin's "The Narrator" to today, noting how "the question of experience can be approached nowadays only with an acknowledgment that it is no longer accessible to us. For just as man has been deprived of his biography, his experience has likewise been expropriated. Indeed, his incapacity to have and communicate experiences is perhaps one of the few self-certainties to which he can lay claim" (13). The incapacity of contemporary man to translate experience is what makes daily existence unbearable, according to Agamben, and not because of any sort of

"alleged poor quality of life or its meaninglessness compared with the past" (14). Giorgio Agamben, *Infancy and History: The Destruction of Experience,* trans. Liz Heron (London: Verso, 1993). It is true that Agamben distances himself from that text in his recent introduction to it, but the term *contemporary* does appear in another text titled, appropriately enough, "What Is the Contemporary?" There, however, contemporaneity takes on a thanatopolitical perspective as well when Agamben, glossing a poem of Osip Mandelstam's, writes that "the poet who must pay his contemporariness with his life, is he who must firmly lock his gaze onto the eyes of the century-beast." Giorgio Agamben, "What Is the Contemporary?," in *What Is an Apparatus?,* 42. Here, too, degrees of contemporaneity are marked by the increasing inscription of death in life.

44 Cited in Agamben, *What Is an Apparatus?,* 2.

45 Gilles Deleuze, "What Is a Dispositif?," in *Michel Foucault Philosopher,* trans. Timothy J. Armstrong (New York: Routledge, 1992), 161.

46 Ibid.

47 Ibid.

48 Michel Foucault, "The Risks of Security," in *Power,* ed. James Faubion, trans. Robert Hurley et al. (New York: New Press, 1997), 369.

49 See, in particular, the appendix to Agamben, *The Kingdom and the Glory.*

50 Agamben, *What Is an Apparatus?,* 18–19.

51 Agamben, *Means without End,* 116.

52 Compare Derrida's reading of separation and religion on this score: "Thus is announced the anchoritic community of those who love in separation. . . . The invitation comes to you from those who can *love only at a distance, in separation.* . . . Those who love only in cutting ties are the uncompromising friends of solitary singularity." Jacques Derrida, *The Politics of Friendship,* trans. George Collins (New York: Verso, 1997), 35; emphasis original.

53 Agamben, *What Is an Apparatus?,* 20.

54 See Michel Foucault, "'Omnes et Singulatim': Towards a Critique of Political Reason," in Faubion, *Power,* esp. 319–22.

55 Michel Foucault, *The Birth of Biopolitics: Lectures at the Collège de France, 1978–1979,* trans. Graham Burchell (New York: Palgrave, 2008), 27.

56 Agamben, *What Is an Apparatus?,* 24.

57 Ibid., 20–21.

58 Ibid., 22.

59 Ibid., 15.

60 Ibid., 14.

61 Ibid.

62 Ibid.

63 Ibid., 19.

64 Agamben, *Means without End,* 116.

65 Agamben, *What Is an Apparatus?,* 11.

66 Ibid., 17–18. This distinction plays out in Agamben, *The Kingdom and the Glory,* as one between an immanent trinity and an economic trinity: "The fracture between being and praxis is marked in the language of the Church Fathers by the terminological opposition between theology and *oikonomia.*" Giorgio Agamben, *Il regno*

e la gloria: Per una genealogia dell'economia e del governo (Vincenza, Italy: Neri Pozza, 2007), 76. See as well 125–39 and 228–30.

67 Heidegger, *Parmenides*, 84.

68 Agamben, *What Is an Apparatus?*, 16.

69 Ibid.

70 Ibid., 12.

71 Heidegger, "Question Concerning Technology," 21.

72 Agamben, *What Is an Apparatus?*, 12.

73 Ibid., 23–24.

74 Ibid., 22.

75 Ibid., 24.

76 Giorgio Agamben, "No to Bio-political Tattooing," *Le monde*, January 10, 2004.

77 Agamben, *The Open*, 77.

78 Michel Foucault, *"Society Must Be Defended": Lectures at the Collège de France, 1975–1976*, trans. David Macey (New York: Picador, 2003), 260.

79 Heidegger, "Letter on Humanism," 199–200.

80 Agamben, *Sacrament of Language*, 91.

81 Ibid., 70.

82 Ibid.

83 Ibid., 71.

84 Ibid.

85 Ibid., 95.

86 Rène Girard, *Violence and the Sacred*, trans. Patrick Gregory (Baltimore: Johns Hopkins University Press, 1979), 24–25.

87 Agamben, *Sacrament of Language*, 72.

88 Agamben, *Il regno e la gloria*, 312.

89 Agamben, *What Is an Apparatus?*, 15.

90 This is not to be thought apart from those troubling pages of Agamben's dedicated to "The Idea of Music," in which he attempts to find an explanation for the lack in modern literature and philosophy of "new epochal sentiments." He gives two reasons, admittedly unhappy ones: first, that "what had been the limit-experience of an intellectual elite became mass experience," and second, even more decisively, "the dizzying loss of authority of private existence and of the private life. Just as we no longer believe in ambience . . . so we no longer expect much from the sentiments that furnish our soul." Giorgio Agamben, *The Idea of Prose*, trans. Michael Sullivan and Sam Whitsitt (Albany: State University of New York Press, 1995), 90.

91 Agamben, *What Is an Apparatus?*, 14.

92 Deleuze, "What Is a Dispositif?," 163.

93 Ibid.

94 Stefano Rodotà's recent work, *La vita e le regole: Tra diritto e non diritto* (Milan, Italy: Feltrinelli, 2006) figures prominently in this revalorization of the category of person in contemporary debates. "'The slipping of attention from the subject to the person, proven by the prevalence of this latter word in the greater part of recent literature' is not born from its superior degree of abstraction, but on the contrary by its greater connection to the material situation of the living individual." Quoted in Esposito, *Terza persona*, 7.

95 With that said, Agamben does take up the impersonal in ways quite different from Esposito in two recent texts. In “Genius,” he relates genius to the impersonal as what does not belong to us: “To have emotion, to be moved, is to feel the impersonal within us, to experience Genius as anguish or joy, safety or fear.” Giorgio Agamben, “Genius,” in *Profanations*, trans. Jeff Fort (New York: Zone Books, 2007), 15. In the second, he links the impersonal to “bare life” as well as to the liberating desire to be free of the guilt and ethics that the personal implies. Giorgio Agamben, *Nudities*, trans. David Kishik and Stefan Pedatella (Palo Alto, Calif.: Stanford University Press, 2011), 52–53.
96 Esposito, *Terza persona*, 6.
97 Esposito, “Dispositif of the Person.”
98 Ibid.
99 See John Caputo, *Heidegger and Aquinas: An Essay on Overcoming Metaphysics* (New York: Fordham University Press, 1982).
100 Saint Augustine, *De Trinitate*, 15: 7–11.
101 Agamben, *What Is an Apparatus?*, 16. See as well the first chapter of Agamben, *Il regno e la gloria*, esp. 24–28.
102 Agamben, *What Is an Apparatus?*, 11.
103 Esposito, “Dispositif of the Person,” 5–6.
104 Ibid., 6.
105 Max Weber, *Theory of Social and Economic Organization* (New York: Free Press, 1997), 241.
106 See, in particular, the works of Catholic philosopher Jacques Maritain, esp. *The Education of Man: The Educational Philosophy of Jacques Maritain* (Garden City, N.Y.: Doubleday, 1962); Maritain, *Integral Humanism* (Notre Dame, Ind.: Notre Dame University Press, 1996); and Maritain, *The Person and the Common Good* (Notre Dame, Ind.: Notre Dame University Press, 1972).
107 Esposito, *Terza persona*, 13–14.
108 Compare Bruno Latour’s recent reading of the thing and the personal: “To put it another way: *objects and subjects can never associate with one another; humans and nonhumans can.* As soon as we stop taking nonhumans as objects, as soon as we allow them to enter the collective in the form of new entities with uncertain boundaries, entities that hesitate, quake, and induce perplexity, it is not hard to see that we can grant them the designation of actors.” Latour, *Politics of Nature: How to Bring the Sciences into Democracy*, trans. Catherine Porter (Cambridge, Mass.: Harvard University Press, 2004), 76.
109 “Therefore I argue . . . that the sorts of knowledge genomics provides allows us to *grammatically* conceive of life in certain ways, *not* in terms of an Aristotelian poesies, but rather as that whose futures we can calculate in terms of probabilities of certain disease events happening—and this shifting grammar of life, towards a future tense, is consequential not just to our understanding of what ‘life’ now means, but contains within it a deep ethical valence.” Kaushik Sunder Rajan, *Biocaptial: The Constitution of Postgenomic Life* (Durham, N.C.: Duke University Press, 2006), 14; emphasis original.
110 Esposito, *Bíos*, 65.
111 See esp. ibid., 135–45.

112 Esposito, *Terza persona*, 17.

113 For a compelling reading of neoliberal trade in organs in India and its relation to the body proper, see Lawrence Cohen, "The Other Kidney: Biopolitics beyond Recognition," *Body and Society* 7 (June–September 2001): 9–29. "It is that the one life cannot subsist any more without the mobilization of the other, that the shoring up of life as it always was—bios, the conservatism of the family form—demands someone else's materiality, *the traffic in zoē*" (25–26; emphasis original).

114 Foucault, *Birth of Biopolitics*, 250–55.

115 Esposito, *Categorie dell'impolitico* (Bologna, Italy: Il Mulino, 1999), 189.

116 Peter Singer, *Practical Ethics* (Cambridge: Cambridge University Press, 1993), 87. See as well Wolfgang Frühwald's interview, "Der 'optimierte' Mensch," in *Biopolitik: Die Positionen*, ed. Christian Geyer, 275–85 (Frankfurt: Suhrkamp, 2001). Responding to the question about the point at which a human individual has worth, he notes, "To such a question there are only individual answers, because there is no firm biological definition of when someone becomes a person" (278).

117 Ulrich Beck and Ulf Erdmann Ziegler, *Eigenes Leben: Ausflüge in die Unbekannte Gesellschaft, in der Wir Leben* (Munich, Germany: C. H. Beck, 1997), 5; cited in Joãh Biehl, "Vita: Life in a Zone of Social Abandonment," *Social Text* 19, no. 3 (2001): 135.

118 Joãh Biehl, "The Activist State: Global Pharmaceuticals, AIDS, and Citizenship in Brazil," *Social Text* 22, no. 3 (2004): 122.

119 Esposito, *Communitas*, 96.

120 Ibid., 99.

121 Ibid., 100.

122 For a brilliant reading of this passage, see Alberto Moreiras, "La vertigine della vita: su *Terza persona* di Roberto Esposito," in *L'impersonale: In dialogo con Roberto Esposito* (Milan, Italy: Mimesis, 2008), 149–72.

123 Weil, "Human Personality," 82; Esposito, *Terza persona*, 125.

124 See in this regard Michael Sandel, *Liberalism and the Limits of Justice* (Cambridge: Cambridge University Press, 1998), as well as the anthology he edited, *Liberalism and Its Critics* (New York: New York University Press, 1984).

125 Martin Heidegger, *Being and Time*, trans. John Macquarrie and Edward Robinson (New York: Harper Perennial Modern Thought, 2000), 164.

126 Heidegger, "Letter on Humanism," 198.

127 Michel Foucault, "The Ethics of the Concern for Self as a Practice of Freedom," in *Ethics: Subjectivity and Truth*, ed. Paul Rabinow, trans. Robert Hurley et al. (New York: New Press, 1997), 282–83.

128 Gilles Deleuze, "On Philosophy," in *Negotiations: 1972–1990*, trans. Martin Joughin (New York: Columbia University Press, 1990), 147.

129 Agamben, *Nudities*, 44.

130 Foucault, "What Is Enlightenment?," 45–46. Compare, on this score, Bruno Bosteels's afterword to *A Leftist Ontology: Beyond Relativism and Identity Politics* (Minneapolis: University of Minnesota Press, 2009) titled "Thinking, Being, Acting, or, On the Uses and Disadvantages of Ontology for Politics," in which Bosteels raises the question of ontology versus a theory of the subject. In the present discussion, we might well want to ask if the thanatopolitical drift of Agamben's

thought does not actually block a more thoroughgoing critique of a theory of technology as neither proper nor improper.

3. Barely Breathing

1 Both essays are collected in Peter Sloterdijk, *Nicht Gerettet: Versuche nach Heidegger* (Frankfurt am Main, Germany: Suhrkamp, 2001). For a helpful summary of the debate, see Andrew Fischer, "Flirting with Fascism—the Sloterdijk Debate," http://www.radicalphilosophy.com/print.asp?editorial_id=10101 (accessed June 10, 2011). See also Gianluca Bonaiuti's perspective on the figure of Sloterdijk in German philosophy, which appears as the introduction to the Italian translation of Peter Sloterdijk, *Il mondo dentro il capitale,* 9–27 (Rome: Meltimi, 2006).

2 See my "Politica, Immunità, Vita: Il pensiero di Roberto Esposito nel dibattito filosofico contemporaneo," in Esposito, *Termini della politica,* esp. 42–58.

3 See primarily Kittler, *Discourse Networks 1800/1900,* as well as Kittler, *Gramophone, Film, Typewriter.*

4 Peter Sloterdijk, *Sphären: Makrosphärologie* (Frankfurt am Main, Germany: Suhrkamp, 1999).

5 Peter Sloterdijk, *Terror from the Air,* trans. Amy Patton and Steve Corcoran (Los Angeles, Calif.: Semiotext(e), 2009).

6 Sloterdijk, *Rage and Time,* 232n9.

7 The complete phrase is "Wohnens oder des Bei-sich-und-den-Seinen-Seins." Sloterdijk, *Sphären,* 996.

8 For a helpful summary of the role of space in Heidegger's thought after the turning in the 1930s, see Malpa, *Heidegger's Topology.* See as well Andrew J. Mitchell's recent volume on Heidegger and sculpture, *Heidegger among the Sculptors: Body, Space, and the Art of Dwelling* (Palo Alto, Calif.: Stanford University Press, 2010).

9 Peter Sloterdijk, "Im Dasein liegt eine wesenhafte Tendenz auf Nähe," in *Nicht gerettet,* 403; quoted in Bonaiuti's introduction to Sloterdijk, *Il mondo dentro il capitale,* 19.

10 Sloterdijk, *Sphären,* 817.

11 Ibid., 812.

12 Ibid.

13 Heidegger, *Elucidations of Hölderlin's Poetry,* 42.

14 Sloterdijk, *Sphären,* 812–13.

15 Heidegger, *Elucidations of Hölderlin's Poetry,* 42.

16 Pier Aldo Rovatti, *Abitare la distanza: Per una pratica della filosofia* (Milan, Italy: Cortina Raffaello, 2007).

17 Peter Sloterdijk, *Im Weltinnenraum des Kapitals* (Frankfurt: Shurkamp, 2007), 244.

18 For a different perspective on contamination, compare Carlo Galli's "Contaminazione: Irruzione del nulla," in *Nichilismo e politica,* 138–58 (Rome: Laterza, 2000).

19 Compare on this point his readings of the modern classics as attempts by their protagonists to conquer the world by devising sufficient protections that will qualify also as self-conquests. Sloterdijk, *Sphären,* 946.

20 See Campbell, "Politics, Immunità, Vita."

21 Sloterdijk, *Sphären,* 948.

22 Ibid.
23 Ibid., 949.
24 Achille Mbembe, "Necropolitics," *Public Culture* 15, no. 1 (2003): 11–40. Of special interest is the affinity between Sloterdijk's reading of immunization and Mbembe's "state of siege": "The *state of siege* is itself a military institution. It allows a modality of killing that does not distinguish between the external and the internal enemy. Entire populations are the target of the sovereign" (30).
25 See Jürgen Habermas, *The Postnational Constellation* (London: Polity Press, 2000).
26 Sloterdijk, *Sphären,* 1003–4.
27 See Bruno Latour and Peter Weibel's edited volume *Making Things Public: Atmospheres of Democracy* (Cambridge, Mass.: MIT Press, 2005), esp. Sloterdijk's contribution "Instant Democracy: The Pneumatic Parliament," 952–57.
28 Sloterdijk, *Sphären,* 950.
29 Ibid., 953.
30 Thomas Hobbes, "On Man," in *Man and Citizen* (Indianapolis, Ind.: Hackett, 1991), 49. See as well Carl Schmitt, *The Leviathan in the State Theory of Thomas Hobbes,* trans. George Schwab and Erna Hilfstein (Westport, Conn.: Greenwood Press, 1996), 34–35: "The leviathan thus becomes none other than a huge machine, a gigantic mechanism in the service of ensuring the physical protection of those governed."
31 Heidegger, *Parmenides,* 87.
32 Sloterdijk, *Sphären,* 1003.
33 Heidegger, *Parmenides,* 84.
34 Sloterdijk, *Sphären,* 1004.
35 Ibid., 994.
36 Ibid., 992.
37 Ibid., 994.
38 Sloterdijk, *Terror from the Air,* 98–99.
39 On immunitary tolerance, see, of course, Donna Haraway, "The Biopolitics of Postmodern Bodies," in *Simians, Cyborgs, and Women: The Reinvention of Nature,* 203–30 (New York: Routledge, 1991). For a recent critique of liberal tolerance, compare Massimo Cacciari's *La maschera della tolleranza.*
40 When combined with the walls of status and immobility, its effects are even more lethal. "Status creates a social scale which determines who may go into areas considered attractive. Those who cannot enter any place are condemned to wander in a 'social and spatial no-man's land. For them, there is only the exterior, the outside.' It is the dumping ground of democratic and liberal inclusion." Oliver Razac, *Barbed Wire: A Political History,* trans. Jonathan Kneight (New York: New Press, 2002), 114.
41 Sloterdijk, *Terror from the Air,* 22–23.
42 Ibid., 25.
43 See Neil A. Lewis, "Broad Use of Harsh Tactics Is Described at Cuba Base," *New York Times,* April 17, 2004.
44 Sloterdijk, *Terror from the Air,* 25.
45 Ibid., 28.

46 Ibid.
47 Ibid., 29.
48 Cf. Jacques Derrida, "Faith and Knowledge: The Two Sources of Religion," in *On Religion,* ed. Jacques Derrida and Gianni Vattimo (Palo Alto, Calif.: Stanford University Press, 1998), and Derrida, "Autoimmunity: Real and Symbolic Suicides," in *Philosophy in the Time of Terror: Dialogues with Jürgen Habermas and Jacques Derrida,* ed. Giovanna Borradori (Chicago: University of Chicago Press, 2003).
49 Esposito, *Communitas,* 19.
50 See Rainer Emig, *Modernism in Poetry: Motivation, Structures, and Limits* (New York: Longman, 1995), 109.
51 Sloterdijk, *Terror from the Air,* 80.
52 Alain Badiou, *The Century,* trans. Albert Toscano (London: Polity, 2007), 160–61.
53 Sloterdijk, *Rage and Time,* 9.
54 Paul Virilio, *Speed and Politics* (New York: Semiotext(e), 1986).
55 Sloterdijk, *Rage and Time,* 122.
56 Ibid.
57 Ibid., 128.
58 Ibid.
59 Ibid., 129.
60 Ibid. Compareon this score Simone Weil's reminder about the end of capitalism: "Since the 'destruction of capitalism' has no meaning—capitalism being an abstraction—and since it does not refer to any precise modifications that might be applied to the regime . . . the slogan can only imply the destruction of capitalists and, more generally, of everyone who does not call himself an opponent of capitalism." Weil, "The Power of Words," in Miles, *Simone Weil: An Anthology,* 231.
61 Sloterdijk, *Rage and Time,* 129.
62 Ibid.
63 Michel Foucault, "*Society Must Be Defended,*" 261. Cf. Ernst Nolte's reading of fascism in *The Three Faces of Fascism* as inscribed within an anti-Marxism: "If fascism is a form of anti-Marxism designed to exterminate its opponent, it cannot be satisfied with the mere political defeat of a party: it must expose the spiritual roots and include them in its condemnation" (51). With regard to Marxism and life, he writes, "Its [Marxism's] transcending nature formed the historic link with the most recent social phenomenon, the still oppressed but already emerging proletariat, thus enabling it to become Europe's most recent faith. . . . It was this apparently monolithic unity which engendered that widespread fear and hatred without which fascism could never have arisen, for in its negation, fascism took this unity with desperate seriousness" (551).
64 Foucault, "*Society Must Be Defended,*" 262.
65 Sloterdijk, *Rage and Time,* 211.
66 Ibid., 212. Compare Sloterdijk's reading of the Weimar zeitgeist: "Every polemical subjectivity arises in the final analysis from the struggles of denial of egos against pain, which they inevitably encounter as living beings. They carry on 'reconstruction,' armament, wall building, fencing in, demarcation, and self-hardening in order to protect themselves." Peter Sloterdijk, *Critique of Cynical Reason,* trans. Michael Eldred (Minneapolis: University of Minnesota Press, 1987), 468.

67 Sloterdijk, *Rage and Time,* 205.

68 Ibid., 225. See Oriana Fallaci, *The Rage and the Pride* (New York: Rizzoli, 2002).

69 Ibid.

70 Heidegger, *Parmenides,* 86.

71 For an introduction to this important thematic, see Michael Hardt, "Introduction: Laboratory Italy," in *Radical Thought in Italy,* ed. Paolo Virno and Michael Hardt, 1–10 (Minneapolis: University of Minnesota Press, 1996). In the same volume, see as well Maurizio Lazzarato, "Immaterial Labor," 133–46.

72 See on this point Heidegger on potentiality and care: "The resoluteness toward itself first brings Da-sein to the possibility of letting the others who are with it 'be' in their own most potentiality-of-being, and also discloses that potentiality in concern which leaps ahead and frees. Resolute Da-sein can become the 'conscience of others'." Heidegger, *Being and Time,* 299; quoted in Esposito, *Communitas,* 95.

73 Sloterdijk, *Sphären,* 995.

74 Foucault, *"Society Must Be Defended,"* 245.

75 For a recent and impressive exception, see Robyn Marasco, "Machiavelli *contra* Governmentality," paper presented at "Biopolitics and Its Vicissitudes," Amherst College, Amherst, Mass., April 2009.

76 Foucault, *"Society Must Be Defended,"* 249. On biopolitics and risk, see Pat O'Malley, "Risk and Responsibility," in *Foucault and Political Reason: Liberalism, Neo-liberalism, and Rationalities of Government,* ed. Andrew Barry, Thomas Osborne, and Nikolas Rose, 189–207 (Chicago: University of Chicago Press, 1997).

77 Sloterdijk does use the term *biopolitics* in *Sphären,* esp. in volume 1, when speaking of the Roman amphitheater. My thanks to David Rojas for pointing out the reference.

78 Foucault, *"Society Must Be Defended,"* 254.

79 Montag, "Necro-economics," in *Radical Philosophy,* 17.

80 Peter Sloterdijk, "Domestikation des Seins: Die Verdeutlichung der Lichtung," in *Nicht gerettet,* 203.

81 Ibid.

82 "The world of racism thus described is a world of incompatible particularities; a world in which a genuine human universalism is as unthinkable as a universalism of life. . . . It is this view that contemporary biology and genetics invalidates with the finding that evolution has produced broad genetic commonality, a vital universality, with genetic variation occurring through mutation primarily at the level of the individual rather than the group." Warren Montag, "Toward a Conception of Racism without Race: Foucault and Contemporary Biopolitics," *Pli* 13 (2002): 121.

83 Sloterdijk, "Domestikation des Seins," 205.

84 Cf. Sloterdijk, "Rules for the Human Zoo," 25–27. See as well Massimo Cacciari's reading of the mystical in Heidegger: "If, on one hand, the dimension of the Mystical as disclosing is similar to the Heideggerian problem of Aletheia, on the other hand, it recalls Rosenzweig's revelation. Disclosing, as Presupposition, is Revelation." Cacciari, *Dalla Steinhof: Prospettive viennesi del primo Novecento* (Milan, Italy: Adelphi, 1980), 139–40.

85 Sloterdijk, "Rules for the Human Zoo," 26. It would appear that Sloterdijk sees modern protocols of biotechnology forgoing any relation whatsoever to prosthetics.

See his discussion of Dessauer and the affirmation of technology, which takes on particular importance in this context: "The fighting subject made of heroism and steel has to be blind to its own destructiveness. The more it threatens to break under the massive suffering of the technical, dominated world, the more optimistically it simulates the heroic pose. At the heart of this theory stands a subject who can no longer suffer because it has become wholly prosthesis." Sloterdijk, *Critique of Cynical Reason,* 457.

86 See Mario Perniola, *Miracoli e traumi della comunicazione* (Turin, Italy: Einaudi, 2009), for another Heideggerian-inflected reading of communication in the Italian context.

87 On the ambivalences that link Nazism to Plato and our own philosophical tradition, see Simona Forti, "The Biopolitics of Souls: Racism, Nazism, and Plato," *Political Theory* 34, no. 1 (2006): 9–32.

88 It might be helpful to read in this defense of bioengineered humans something like Sloterdijk's earlier reading of Dionysian learning, what he calls alternately "'therapeutics,' 'psychonautics,' or 'psychodrama,'—yes even 'politics.'" Thus "a therapeutic drama at the level of universal civilization, which would be carried out without anyone authorizing or ordering it, would be a learning process that could bring to an end the assault of active nihilism, with its assignments of value, constructive measures, establishment of levels, and eliminations." Peter Sloterdijk, *Thinker on Stage: Nietzsche's Materialism,* trans. Jamie Owen Daniel (Minneapolis: University of Minnesota Press, 1989), 89.

4. Practicing *Bíos*

1 Bernard Stiegler, *Technics and Time: The Fault of Epimetheus, 1,* trans. Richard Beardsworth and George Collins (Palo Alto, Calif.: Stanford University Press, 1998), 187.

2 "One might say that the ancient right to take life and let live was replaced by a power to foster life or disallow it to the point of death." Michel Foucault, *History of Sexuality Volume 1: An Introduction,* trans. Robert Hurley (New York: Vintage Books, 1978), 138. See as well my introduction to Roberto Esposito, *Bíos.* I also take up these issues with Adam Sitze in our introduction to *Biopolitics: A Reader* (Durham, N.C.: Duke University Press, forthcoming).

3 It is true that Foucault will also make biopolitics the object of his following talk from 1979 titled *The Birth of Biopolitics,* and there, too, he will observe how "the central core of all the problems that I am trying to identify is what is called the population" and that "this is the basis on which something like biopolitics could be formed." Foucault, *Birth of Biopolitics,* 21. For the most part, however, he dedicates his analysis to the similarities between ordoliberalism in Germany and neoliberalism in the United States and forgoes an analysis of the role technology plays in both.

4 Foucault, *Security, Territory, Population,* 1.

5 See Jean-Baptiste Lamarck, *Zoological Philosophy: An Exposition with Regard to the Natural History of Animals,* trans. Hugh Eliot (London: Macmillan, 1914). A helpful summary of Lamarck's notion of *milieux* can be found in Margo Huxley, "Spatial Rationalities: Order, Environment, Evolution, and Government," *Social and Cultural Geography* 7, no. 5 (2006): 771–87.

6 Foucault, *Security, Territory, Population*, 42.

7 Foucault had introduced the question of security and biopower earlier in *"Society Must Be Defended."* Speaking of the emergence of biopower, Foucault writes, "In a word, security mechanisms have to be installed around the random element inherent in a population of living beings so as to optimize a state of life." Foucault, "*Society Must Be Defended,*" 246. Compare, on this score, the role of fear in the current war on terror as well as the financial derivatives trade: "Terror evinces a politics of fear as well as vulgarity, but we should not cede to the state the affective monotone that would be most conducive to its effective rule. . . . The move from uncertainty to risk is a shift from the fear of the unknown to the thrill of the unexpected." Randy Martin, *An Empire of Indifference* (Durham, N.C.: Duke University Press, 2007), 142.

8 Foucault, *Security, Territory, Population*, 42. Warren Montag registers this even more forcefully than Foucault in his "Necro-Economics: Adam Smith and Death in the Life of the Universal," *Radical Philosophy* 134 (November–December 2005): 7–17: "If societies, by virtue of the economy of nature, must exercise, and not merely possess, the right to kill, the market, understood as the very form of human universality as life, must necessarily, at certain precise moments, 'let die'" (13).

9 The strengthening of individual national identities is one response to a scarcity of security. However, "once the problem of invisible contagion is admitted, it is immediately answered by a further specification of the collapse between national and physical borders: invisibility is made visible through signifiers of national and ethnic identity." Kirsten Ostherr, *Cinematic Prophylaxis: Globalization and Contagion in the Discourse of World Health* (Durham, N.C.: Duke University Press, 2005), 73.

10 See, too, the "new subject of the times" after World War I in Sloterdijk, *Critique of Cynical Reason*, 434, as well as Jonathan Crary's *Techniques of the Observer: On Vision and Modernity in the 19th Century* (Cambridge, Mass.: MIT Press, 1992), esp. 18–21.

11 Foucault, *Security, Territory, Population*, 44. See in this regard Nico Poulantzas's gloss of resistance and law in *State, Power, Socialism*, trans. Patrick Camiller (London: Verso, 2000), 76–92, 149–50. "Hence Foucault's problem: how is it possible to avoid falling into the conceptual trap of a domination that cannot be escaped; of a power that is absolutely privileged in relation to resistances; of resistances that are always ensnared by power? There can only be one answer: it is necessary to break loose from this hypostasized power and rediscover at any cost something other than these resistances inscribed in power" (150).

12 Nor does Agamben in those pages dedicated to people/People in *Means without End*, in which he elides the passage from multiplicity to people in Foucault. See esp. part I, "What Is a People?," 29–36.

13 Foucault, *Security, Territory, Population*, 45.

14 Ibid. Compare in this regard Antonello Petrillo's entry for "Sicurezza" in *Lessico di biopolitica*, ed. R. Brandimarte, P. Chiantera-Stutte, P. di Vittorio, O. Marzocca, O. Romano, A. Russo, and A. Simone (Rome: Manifestolibri, 2006), 289–94.

15 And hence linked to biopower: "'The history of biopower teaches us that the paradigm of security on which contemporary *dispositifs* of control are based functions

by projecting a specter consisting of diverse fears: to govern means to manage desires by animating fears; to separate therefore the phantasmorgia of a desirable existence from our lives, or, and which amounts to the same, of an existence to be feared.' And thus 'every security is conquered by an insecurity and generates in turn a new insecurity' (Helmut Plessner, *Macht und menschliche Natur*, 1931)." Quoted in Andrea Cavalletti, *Classe* (Torino, Italy: Bollati Boringhieri, 2009), 90–91.

16 Foucault, *Security, Territory, Population*, 49.

17 Ibid. On the relation of personhood for biopolitics, see Bryn Williams-Jones, "Concepts of Personhood and Commodification of the Body," *Health Law Review* 7, no. 3 (1999): 11–13. See also Mark J. Hanson, ed., *Claiming Power over Life: Religion and Biotechnology Policy* (Washington, D.C.: Georgetown University Press, 2001). In a recent volume of *Micromega*, two Italian philosophers take up the problematic of personhood. See Roberto Esposito and Stefano Rodotà, "La maschera della persona," *Micromega* 3 (2007): 105–15.

18 Aihwa Ong, *Neoliberalism as Exception: Mutations in Citizenship and Sovereignty* (Durham, N.C.: Duke University Press, 2006), 185. Continuing, Ong adds, "The ethical questions are of what it means to be cultural beings are now framed within the transnational artificial environments dominated by the interests of global pharmaceutical firms."

19 On the history of eugenics, see Daniel Kevles, *In the Name of Eugenics: Genetics and the Uses of Human Heredity* (Cambridge, Mass.: Harvard University Press, 1995). For a brilliant exposition of Agamben's thought in relation to eugenics, see Catherine Mills, "Biopolitics, Liberal Eugenics, and Nihilism," in *Giorgio Agamben: Sovereignty and Life*, 180–202.

20 Foucault, *Security, Territory, Population*, 16.

21 Massimo Cacciari, "Nomadi in prigione," in *La città infinita*, ed. A. Bonomi and A. Abruzzese (Milan, Italy: Bruno Mondadori, 2004), 45. Unless otherwise noted, all translations are my own.

22 "Our point of departure is our recognition that the production of subjectivity and the production of the common can together form a spiral, symbiotic relationship. . . . Perhaps in this process of metamorphosis and constitution we should recognize the formation of the body of the multitude, a fundamentally new kind of body, a common body, a democratic body." Michael Hardt and Antonio Negri, *Multitude: War and Democracy in the Age of Empire* (New York: Penguin, 2005), 189–90. Compare Warren Montag on the figure of the multitude: "Spinoza, in his search for stability and political equilibrium, turns a juridical formalism against the power of the multitude. . . . The alternative . . . is daunting indeed; it is we might say a politics of permanent revolution, a politics utterly without guarantees of any kind, in which social stability must always be re-created through a constant reorganization of corporeal life, by means of a perpetual mass mobilization." Warren Montag, *Bodies, Masses, Power: Spinoza and His Contemporaries* (New York: Verso, 1999), 84–85.

23 Foucault, *Birth of Biopolitics*. Warren Montag also points to this mechanism in his introduction of the term *necroeconomics* as homonymous with the market and the process whereby some are slowly killed so as to make the market work. "Smith's economics is a necro-economics. The market reduces and rations life; it not only

allows death, it demands that death be allowed by the sovereign power, as well as by those who suffer it." Montag, "Necro-economics," 16.

24 Or more simply, populations of individuals make the market work, populations that may be allowed to die. Again, Montag's perspective comes into view: "The subsistence of a population may, and does in specific circumstances, require the death of a significant number of individuals, to be precise it requires that they be allowed to die so that others may live." Montag, "Necro-economics," 14. According to this reading, as in Foucault's reading of ordoliberalism and neoliberalism, the guarantee of competition, which is what *govern* for the market means, presumes that some are not competitive and hence will not be "governed."

25 Foucault, *Hermeneutics of the Subject*, 9.

26 "The critical ontology of ourselves must be considered not, certainly, as a theory, a doctrine, nor even as a permanent body of knowledge that is accumulating; it must be conceived as an attitude, an ethos, a philosophical life in which the critique of what we are is at one and the same time the historical analysis of the limits imposed on us and an experiment with the possibility of going beyond them." Foucault, "What Is Enlightenment?," 319.

27 Ibid.

28 Foucault, *Hermeneutics of the Subject*, 117.

29 *OED Online*, s.v. "belonging."

30 Foucault, *Hermeneutics of the Subject*, 357.

31 Ibid., 361.

32 Ibid., 424.

33 Ibid., 429.

34 One ought to keep in mind Derrida's critique of Foucault's *History of Sexuality Volume 1*, especially with regard to mastery and thanatos: "The same strategy, a strategy profoundly without defense, a strategy that carries with it its own principle of ruin—here is that which *also* problematizes, in its greatest radicality, the instance of power in its mastery. In a short and difficult passage, Freud comes fully to nominate when not to identify a *drive for power* and a *drive for mastery*." Jacques Derrida, "Al di là del principio di potere," in *Essere giusti con Freud: La storia della follia nell'età della psicoanalisi* (Milan, Italy: Raffaello Cortina, 1994), 110. My thanks to Lorenzo Fabbri for drawing my attention to this text.

35 Foucault, *Hermeneutics of the Subject*, 448.

36 Ibid., 431.

37 Ibid., 448.

38 Ibid.

39 Ibid.

40 Foucault picks up this definition in the earlier seminar on subjectivity and truth from the lectures of 1980–81. See the important note to *Hermeneutics of the Subject*, 489n28: "It is the second lecture of the 1981 Collège de France course that Foucault distinguishes between *zoē* (life as the property of organisms) and *bios* (existence as the object of techniques)."

41 Ibid., 486.

42 Such a reading of *technē* as not limited to the test recalls a recent reading of Wittgenstein: "Our practices are thus not exhausted by the idea of a rule. On the

contrary, one thing that Wittgenstein is aimed to show . . . is that one hasn't said particularly much about a practice (such as, for instance, language) when one has simply said that it is governed by rules. In reality, discussion of rules is distorted by the (philosophical) idea of an explanatory or justificatory power in the concept of a rule—an idea that leads to conformity." Sandra Laugier, "Wittgenstein and Cavell: Anthropology, Skepticism, and Politics," in *Claim to Community: Essays on Stanley Cavell and Political Philosophy*, ed. Andrew Norris (Palo Alto, Calif.: Stanford University Press, 2006), 36–37. By adopting practice and not test, we begin to move away from the care for the self that begins with the withdrawal of *technē* to the degree that practice cannot be limited to mere rules or the homology between rules and tests.

43 In another note contained in *Hermeneutics of the Subject*, the editor observes that "the 1981 courses continue to focus exclusively on the status of the aphrodisia in pagan ethics of the first two centuries A.D., while maintaining that we cannot speak of subjectivity in the Greek world, the ethical element being determined as *bios* (mode of life)" (20). Foucault appears to suggest that *bíos* is to be thought apart in Greek thought not only from subjectivity but equally from a care for the self. Such a reading will likely be born out by the forthcoming publication of the 1981 course, "Subjectivité et verité."

44 Michel Foucault, "On the Genealogy of Ethics," in Rabinow, *Ethics: Subjectivity and Truth*, 260.

45 Foucault, *Hermeneutics of the Subject*, 424.

46 Ibid.

47 Foucault, "On the Genealogy of Ethics," 256.

48 Ibid.

49 We must also note that Foucault does not deny an aesthetic component with regard to the Stoics such that both the *technē* of the self and a *technē* of *bíos* share a lack of normalizing thought: "I don't think one can find any normalization in, for instance, the Stoic ethics. The reason is, I think, that the principal aim, the principal target of this kind of ethics, was an aesthetic one." Foucault, "On the Genealogy of Ethics," 254. Compare Foucault's broadside against modern ethics in *The Order of Things: An Archaeology of the Human Sciences* (London: Routledge, 2001), 327–28.

50 Equally, such a perspective of *bíos* as material for aesthetic production puts some distance between Foucault's understanding of *bíos* and Agamben's distinction between *bíos* and *zoē* because Foucault, who was well aware of the distinction between *bíos* and *zoē*, does not make *zoē* the object of aesthetics but rather of *bíos*.

51 Michel Foucault, *Fearless Speech* (New York: Semiotext(e), 2001), 166.

52 Donna Haraway, "Biopolitics of Postmodern Bodies," 224.

53 Judith Butler, "Giving an Account of Oneself," *diacritics* 31, no. 4 (2001): 39.

54 Clearly both Butler and Haraway have Foucault's critique of modernity's missing ethics in mind: "Superficially, one might say that knowledge of man, unlike the sciences of nature, is always linked, even in its vaguest form, to ethics or politics; more fundamentally, modern thought is advancing towards that region where man's Other must become the Same as himself." Foucault, *Order of Things*, 328. Agamben, for his part, is less interested in marking such a complicity than he is

in capturing the self in his scheme of desubjectification and hence in refusing to disconnect the self's capabilities from biopower.

55 I am indebted to the attention that both Leo Bersani and Adam Phillips pay to these two essays in their recent work *Intimacies* (Chicago: University of Chicago Press, 2008). My own reading, though indebted to theirs, departs in the emphasis I place on attention and play.

56 Sigmund Freud, "Drives and Their Fates," in *The Unconscious*, trans. Graham Franklin (London: Penguin, 2005), 27.

57 In this regard, see the entry for "negation" in Jean LaPlanche and J. B. Pontalis, *The Language of Psychoanalysis*, trans. Donald Nicholson-Smith (New York: W. W. Norton, 1974).

58 Freud, "Drives and Their Fates," 29.

59 Sigmund Freud, "Negation," *Standard Edition* XIX (1976): 237.

60 See on this note the life and death instincts in Sigmund Freud's *An Outline of Psychoanalysis*, trans. James Strachey (New York: W. W. Norton, 1949), 20: "After long doubts and vacillations we have decided to assume the existence of only two basic instincts, Eros and the destructive instinct. . . . The aim of the first of these basic instincts is to establish ever greater unities and to preserve them thus—in short to bind them together; the aim of the second, on the contrary, is to undo connections and so to destroy things. We may suppose that the final aim of the destructive instinct is to reduce living things to an inorganic state."

61 Freud, "Negation," 239.

62 Freud, "Drives and Their Fates," 29.

63 Useful in this regard is Paolo Virno's recent discussion of the negation of a negation that empowers the force of the *katechon*. Virno, *Multitude between Innovation and Negation*, trans. Isabella Bertoletti, James Cascaito, and Andrea Casson (New York: Semiotext(e), 2008), 56–65.

64 Foucault, *Fearless Speech*, 166.

65 See in this regard Hal Foster's classic reading of Freud and modernity, "Prosthetic Gods," *Modernism/Modernity* 4, no. 2 (1997): 5–38.

66 The Invisible Committee, *The Coming Insurrection* (New York: Semiotext(e), 2009), 29–30.

67 Pierre Bourdieu, *Outline of a Theory of Practice*, trans. Richard Nice (Cambridge: Cambridge University Press, 1977), 2. Compare Foucault's understanding of practices: "In this piece of research on the prisons . . . the target of analysis wasn't 'institutions,' 'theories,' or 'ideology,' but *practices*—with the aim of grasping the conditions which make these acceptable at a given moment; the hypothesis being that these types of practices are not just governed by institutions, prescribed by ideologies, guided by pragmatic circumstances . . . but possess up to a point their own specific regularities, logic, strategy, self-evidence, and reason." Michel Foucault, "Questions of Method," in *The Foucault Effect: Studies in Governmentality*, ed. Graham Burchell, Colin Gordon, and Peter Miller (Chicago: University of Chicago Press, 1991), 75.

68 Bourdieu will also remind us of the impact of different objects on the notion of practice: "Knowledge does not merely depend, as an elementary relativism teaches, on the particular standpoint an observer 'situated in space and time'

takes up on the object. The 'knowing subject' as the idealist tradition rightly calls him, inflicts on practice a much more fundamental and pernicious alteration which . . . is bound to pass unnoticed: in taking up a point of view on the action, withdrawing from it in order to observe it from above and from a distance he constitutes practical activity as an *object of observation and analysis, a representation.*" Bourdieu, *Outline of a Theory of Practice,* 2.

69 Foucault, "What Is Enlightenment?," 314–15.

70 Ibid., 317. "What is at stake, then, is this: how can the growth of capabilities [*capacités*] be disconnected from the intensification of power relations."

71 Crary, *Techniques of the Observer,* 18.

72 Jonathan Crary, *Suspensions of Perception: Attention, Spectacle, and Modern Culture* (Cambridge, Mass.: MIT Press, 1999), 13–14.

73 Maurice Merleau-Ponty, "Attention and Judgment," in *Phenomenology of Perception,* trans. Colin Smith (London: Routledge, 1962), 30.

74 Ibid., 31.

75 Ibid.

76 Ibid., 34.

77 Compare Francis Bacon's perspective on clichés: "It would be much better to abandon oneself to clichés, to collect them, accumulate them, multiply them, as so many prepictorial givens: 'the will to lose the will' comes first." David Sylvester, *The Brutality of Fact: Interviews with Francis Bacon, 1962–1979,* 3rd ed. (New York: Thames and Hudson, 1987), 13; quoted in Gilles Deleuze, *Francis Bacon: The Logic of Sensation,* trans. Daniel W. Smith (Minneapolis: University of Minnesota Press, 2003), 76.

78 Merleau-Ponty, "Attention and Judgment," 35.

79 Deleuze's reading of art in general is apropos here: "Art is defined, then, as an impersonal process in which the work is composed somewhat like a *cairn.* . . . Only a conception such as this can tear art away from the personal process of memory and the collective ideal of commemoration." Gilles Deleuze, "What Children Say," in *Essays Critical and Clinical,* trans. Daniel W. Smith and Michael A. Greco (Minneapolis: University of Minnesota Press, 1997), 66.

80 Merleau-Ponty, "Attention and Judgment," 30.

81 Gilles Deleuze and Félix Guattari, *A Thousand Plateaus: Capitalism and Schizophrenia,* trans. Brian Massumi (Minneapolis: University of Minnesota Press, 1987), 262.

82 Ibid., 263.

83 Deleuze, *Francis Bacon,* 58.

84 *OED Online,* s.v. "attention."

85 Gilles Deleuze, *Cinema 1: The Movement Image,* trans. Hugh Tomlinson and Barbara Habberjam (Minneapolis: University of Minnesota Press, 1986), 115.

86 Jacques Rancière, *Film Fables,* trans. Emiliano Battista (New York: Berg, 2006), 120.

87 Félix Guattari, *Chaosmosis: An Ethico-Aesthetic Paradigm,* trans. Paul Bains and Julian Pefanis (Bloomington: Indiana University Press, 1995), 106.

88 Ibid., 101.

89 Ibid., 102.

90 Ibid., 106.

91 Ibid., 109.

92 Compare the following: "The cut of the present refers to an emptiness that opens a distance between the present and me, between my question about the present and me. It is this distance that allows one to interrogate, for example, the meaning of feminine experience." Chiara Zamboni, "L'inaudito," in *Diotima: Mettere al mondo* (Milan, Italy: La Tartaruga, 1990), 2.

93 Guattari, *Chaosmosis,* 106.

94 Walter Benjamin, "The Concept of Criticism," in *Selected Writings Volume 1 1913–1926* (Cambridge, Mass.: Harvard University Press, 1996), 159.

95 A useful place to start in any archaeology of play will be found in Hans-George Gadamer, *Truth and Method* (New York: Crossroad, 1985), esp. the chapter titled "Play as the Clue to Ontological Explanation." See, in particular, his reading of Kant and Schiller with regard to the former's critique of taste (90–119).

96 George Bataille, "Unknowing and Rebellion," in *The Bataille Reader* (London: Wiley-Blackwell, 1997), 328.

97 Ibid., 327.

98 On play and economy in Bataille's thought, see Arkady Plotnitsky, "The Maze of Taste: On Bataille, Derrida, and Kant," in *On Bataille: Critical Essays,* ed. Leslie Anne Boldt-Irons, 107–28 (Albany: SUNY Press, 1995).

99 Bataille, "Unknowing and Rebellion," 328.

100 Ibid.

101 "He rehabilitates the tangible and human activity of the self but only in order to denounce the illusions it fosters. He insists upon the unity of the human spirit, but in order to rediscover the sacrifice therein and the 'self—for death.' He proclaims love and fusion, but for their relatedness to death." Julia Kristeva, "Bataille, Experience, and Practice," in Boldt-Irons, *On Bataille: Critical Essays,* 239.

102 Jacques Derrida, "Structure, Sign, and Play in the Discourse of the Human Sciences," in *Writing and Difference,* trans. Alan Bass (Chicago: University of Chicago Press, 1978), 292.

103 Ibid.

104 Ibid., 292.

105 Ibid.

106 Ibid.

107 Ibid.

108 Gilles Deleuze, *Spinoza: Practical Philosophy,* trans. Robert Hurley (New York: City Lights, 2001), 94. For a sketch of Deleuze's reading of Spinoza, see Cesare Casarino and Antonio Negri, *In Praise of the Common: A Conversation on Philosophy and Politics* (Minneapolis: University of Minnesota Press, 2008), esp. 31–37.

109 Gilles Deleuze, *Spinoza: Practical Philosophy,* trans. Robert Hurley (San Francisco: City Lights, 1988), 95.

110 In a much later text, Derrida comes closer to what I have in mind with an unsure form of play. When speaking about giving the other "equal time to speak," Derrida writes, "It is not only a matter of letting the other speak, but of letting time speak, its time, what its time, the time of the other, has as its most proper. . . . In letting arrive what arrives (of the other), this letting 'neutralizes' nothing, it is not a simple passivity, even if some passivity is required here. . . . What I would make happen

instead of letting happen, well, that would no longer happen. What I make happen does not happen." Jacques Derrida, *Sovereignties in Question: The Poetics of Paul Celan* (New York: Fordham University Press, 2005), 121.

111 Gadamer knows this well: "The subject of the experience of art, that which remains and endures, is not the subjectivity of the person who experiences it, but the work itself. This is the point at which the mode of being of play becomes significant. For play has its own essence, independent of the consciousness of those who play. Play also exists—indeed exists properly—when the thematic horizon is not limited by any being-for-itself of subjectivity, and where there are no subjects who are behaving 'playfully.'" Hans-Georg Gadamer, *Truth and Method,* trans. Garrett Barden and John Cumming (New York: Seabury Press, 1975), 92. Compare, on this note, Agamben's reading of play in terms of the sacred and profanation: "Just as the *religio* that is played with but no longer observed opens the gate to use, so the power [*potenze*] of economics, law, and politics deactivated in play, can become the gateways to a new happiness." Giorgio Agamben, "In Praise of Profanation," in Fort, *Profanations,* 76. See as well Agamben's *Infancy and History: The Destruction of Experience*, trans. Liz Heron (London: Verso, 1993).

112 For a general overview of Winnicott's thought, see Adam Phillips's indispensable *Winnicott* (Cambridge, Mass.: Harvard University Press, 1988) as well the recent biography by F. Robert Rodman, *Winnicott: Life and Work* (Cambridge, Mass.: Perseus, 2003).

113 D. W. Winnicott, *Playing and Reality* (London: Routledge, 1991), 55.

114 Ibid., 13.

115 Ibid., 14.

116 Ibid., 50. Compare Eric L. Santner's reading of trauma and knowledge: "The recovery of traumatic disruption cannot be conceived as a form of memory (of a scene or event). It involves, rather, the opening to a certain meaninglessness or non-sense—an irrationality—at the heart of the repetition compulsions informing one's way of being in the world and therewith the possibility of changing direction in life. We might say that the mode of verification of a trauma is not some form of recovered memory—some form of historical knowledge—but rather a way of acknowledging a distinctive automaticity at the core of one's being." Santner, *The Psychotheology of Everyday Life: Reflections on Freud and Rosenzeig* (Chicago: University of Chicago Press, 2001), 40. If we were to think play in Winnicott's terms with Santner's, we might find that play would share with a mode of acknowledging a reaching out.

117 *OED Online,* s.v. "attention."

118 Walter Benjamin, "The Cultural History of Toys," in *Selected Writings Volume 2 1927–1934,* ed. Michael W. Jennings (Cambridge, Mass.: Harvard University Press, 1999), 115.

119 My thanks to Kevin Attell for drawing my attention to these forms of play, especially in the thought of Agamben and Gadamer.

120 Walter Benjamin, "Old Toys," in *Selected Writings Volume 2 1927–1934,* 101.

121 William Connolly, "Discipline, Politics, Ambiguity," in *The Self and the Political Order,* ed. Tracy B. Strong (New York: New York University Press, 1992), 159.

122 Friedrich Nietzsche, *The Will to Power*, trans. Anthony M. Ludovici (New York: Barnes and Noble, 2006), 153–54; quoted in Christopher Cox, "The Subject of Nietzsche's Perspectivism," *Journal of the History of Philosophy* 35, no. 2 (1997): 286. The German may be found at *Nietzsches Werke: Historisch-kritische Ausgabe* VII-2 (electronic edition) [*Nachgelassene Fragmente Frühjahr bis Herbst 1884*]: 180–82, http://www.nlx.com/collections/89 (accessed June 10, 2011). On planetary possibilities, see Glissant's "planetary adventure" in terms of relationality. Édouard Glissant, *Poetics of Relation*, trans. Betsy Wing (Ann Arbor: University of Michigan Press, 1997), 162–63.

123 Cf. Bataille's notion of *ipse*: "The *ipse*, the Bataillian self, if we can call it that, is not a high point, not a closed entity embodying a definitive consciousness, but instead a momentary conjunction of coordinated and competing forces, an intersection point, a contingent space of energetic communication." Allan Stoekl, *Bataille's Peak: Energy, Religion, and Sustainability* (Minneapolis: University of Minnesota Press, 2007), 81.

124 Edward F. McGushin, *Foucault's Askēsis: An Introduction to the Philosophical Life* (Evanston, Ill.: Northwestern University Press, 2007), 238. McGushin's larger reading echoes many of the themes of this chapter: "In fact, the main function of biopolitics is to institute this mode of care of the self: it is through this definition of care of the self that individuals are able to be produced and controlled; it is within this system of actuality that they will be confined" (239). The point would be to move toward singularity but with the following proviso: "that singularity, beyond all the theses constructed around its potential for liberty and liberation, can take on a meaning related to forms of opening and vitality only if it is not translated into monadism; only if it refuses to put itself in relation by activating proprietary mechanisms with respect to its own proper self" (299).

125 Such self-knowledge is deeply impolitical, as Sloterdijk suggests. Writing with regard to Nietzsche's *Untimely Meditations*, he notes, "I call this remarkably negative structure of self-knowledge the *psychoautical circle*: Nietzsche's theatrical adventure into the theory of knowledge is intrinsically implicated in it. His personal and philosophical fate depends to a great extent, I believe, on whether he can complete the tasks of burning away images and whether his search for self can be successfully completed within the context of a beneficial negativity and lack of representation." Peter Sloterdijk, *Thinker on Stage: Nietzsche's Materialism*, trans. Jamie Owen Daniel (Minneapolis: University of Minnesota Press, 1989), 34.

126 Compare Kristeva's reading of the self in Bataille: "He rehabilitates the tangible and human activity of the *self* but only in order to denounce the illusions it fosters." Kristeva, "Bataille, Experience, and Practice," 239. Compare also to Bataille's *Theory of Religion*: "It is the most necessary renunciation in one sense: insofar as man ties himself entirely to the real order, insofar as he limits himself to planning operations. But it is not a question of showing the powerlessness of the man of works; it is a question of tearing *man* away from the order of works." George Bataille, *Theory of Religion*, trans. Robert Hurley (New York: Zone Books, 1992), 89.

INDEX

(continued from page ii)

15 *Animal Stories: Narrating across Species Lines*
Susan McHugh

14 *Human Error: Species-Being and Media Machines*
Dominic Pettman

13 *Junkware*
Thierry Bardini

12 *A Foray into the Worlds of Animals and Humans,* with *A Theory of Meaning*
Jakob von Uexküll

11 *Insect Media: An Archaeology of Animals and Technology*
Jussi Parikka

10 *Cosmopolitics II*
Isabelle Stengers

9 *Cosmopolitics I*
Isabelle Stengers

8 *What Is Posthumanism?*
Cary Wolfe

7 *Political Affect: Connecting the Social and the Somatic*
John Protevi

6 *Animal Capital: Rendering Life in Biopolitical Times*
Nicole Shukin

5 *Dorsality: Thinking Back through Technology and Politics*
David Wills

4 *Bíos: Biopolitics and Philosophy*
Roberto Esposito

3 *When Species Meet*
Donna J. Haraway

2 *The Poetics of DNA*
Judith Roof

1 *The Parasite*
Michel Serres

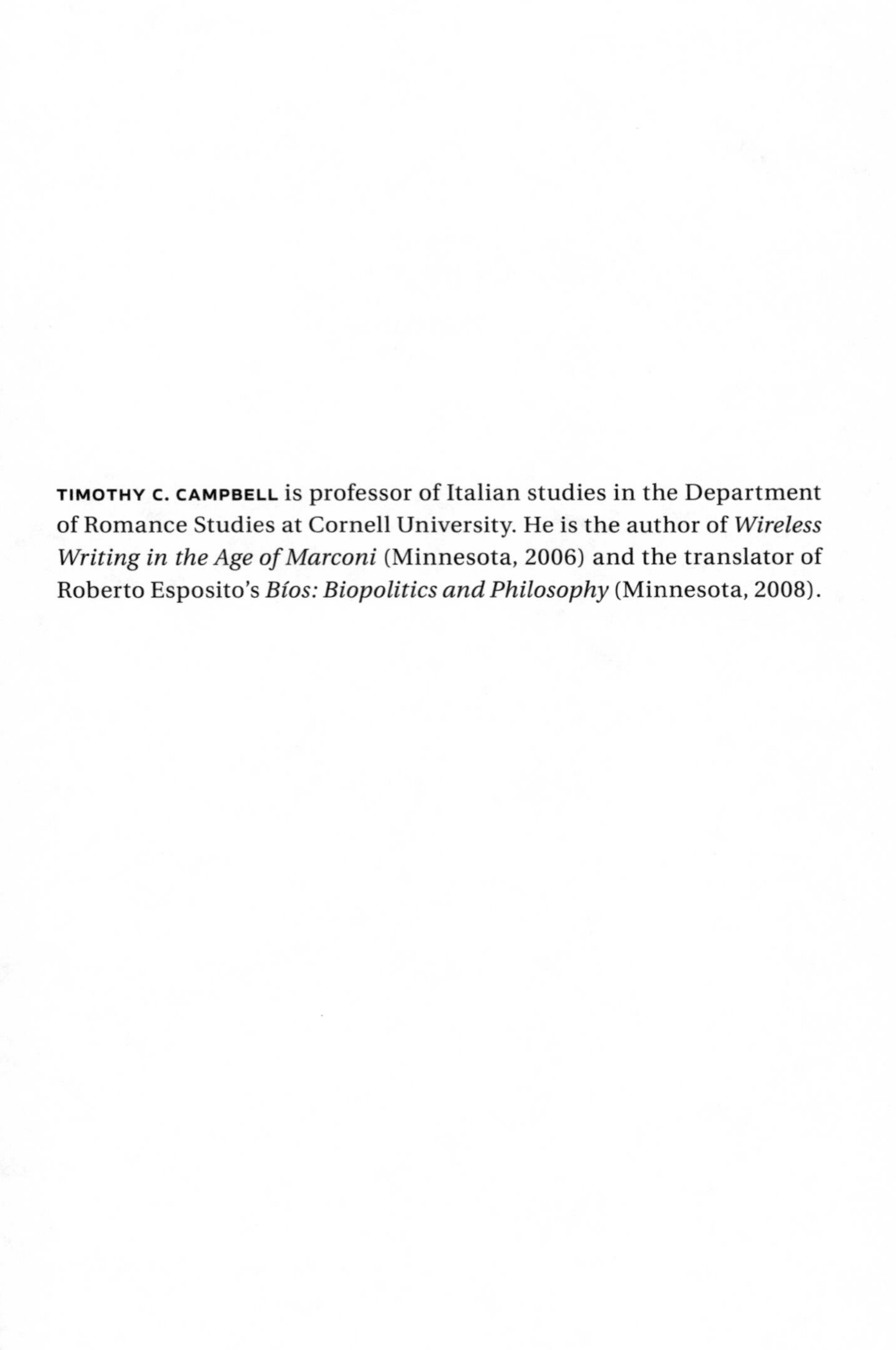

TIMOTHY C. CAMPBELL is professor of Italian studies in the Department of Romance Studies at Cornell University. He is the author of *Wireless Writing in the Age of Marconi* (Minnesota, 2006) and the translator of Roberto Esposito's *Bíos: Biopolitics and Philosophy* (Minnesota, 2008).